T0215803

Data through Movement

Synthesis Lectures on Visualization

Editors

Niklas Elmqvist, *University of Maryland*
David S. Ebert, *University of Oklahoma*

Synthesis Lectures on Visualization publishes 50- to 100-page publications on topics pertaining to scientific visualization, information visualization, and visual analytics. Potential topics include, but are not limited to: scientific, information, and medical visualization; visual analytics, applications of visualization and analysis; mathematical foundations of visualization and analytics; interaction, cognition, and perception related to visualization and analytics; data integration, analysis, and visualization; new applications of visualization and analysis; knowledge discovery management and representation systems and evaluation; distributed and collaborative visualization and analysis.

© Springer Nature Switzerland AG 2022
Reprint of original edition © Morgan & Claypool 2021

All rights reserved. No part of this publication may be reproduced, stored in a retrieval system, or transmitted in any form or by any means—electronic, mechanical, photocopy, recording, or any other—except for brief quotations in printed reviews, without the prior permission of the publisher.

Data through Movement: Designing Embodied Human-Data Interaction for Informal Learning
Francesco Cafaro and Jessica Roberts

ISBN: 978-3-031-01482-6 paperback
ISBN: 978-3-031-02610-2 ebook
ISBN: 978-3-031-00354-7 hardcover

DOI 10.1007/978-3-031-02610-2

A Publication in the Springer series
SYNTHESIS LECTURES ON VISUALIZATION

Lecture #12
Series Editors: Niklas Elmqvist, *University of Maryland*
David S. Ebert, *University of Oklahoma*

Series ISSN
Print 2159-516X   Electronic 2159-5178

Data through Movement
Designing Embodied Human–Data Interaction for Informal Learning

Francesco Cafaro
School of Informatics and Computing, Indiana University-Purdue University Indianapolis

Jessica Roberts
School of Interactive Computing, Georgia Institute of Technology

SYNTHESIS LECTURES ON VISUALIZATION #12

ABSTRACT

When you picture human-data interactions (HDI), what comes to mind? The datafication of modern life, along with open data initiatives advocating for transparency and access to current and historical datasets, has fundamentally transformed when, where, and how people encounter data. People now rely on data to make decisions, understand current events, and interpret the world. We frequently employ graphs, maps, and other spatialized forms to aid data interpretation, yet the familiarity of these displays causes us to forget that even basic representations are complex, challenging inscriptions and are not neutral; they are based on representational choices that impact how and what they communicate. This book draws on frameworks from the learning sciences, visualization, and human-computer interaction to explore embodied HDI. This exciting sub-field of interaction design is based on the premise that every day we produce and have access to quintillions of bytes of data, the exploration and analysis of which are no longer confined within the walls of research laboratories. This volume examines how humans interact with these data in informal (not work or school) environments, paritcularly in museums.

The first half of the book provides an overview of the multi-disciplinary, theoretical foundations of HDI (in particular, embodied cognition, conceptual metaphor theory, embodied interaction, and embodied learning) and reviews socio-technical theories relevant for designing HDI installations to support informal learning. The second half of the book describes strategies for engaging museum visitors with interactive data visualizations, presents methodologies that can inform the design of hand gestures and body movements for embodied installations, and discusses how HDI can facilitate people's sensemaking about data.

This cross-disciplinary book is intended as a resource for students and early-career researchers in human-computer interaction and the learning sciences, as well as for more senior researchers and museum practitioners who want to quickly familiarize themselves with HDI.

KEYWORDS

Human-Data Interaction, Embodied Interaction, Museums, Informal Learning, Data Exploration, Mid-Air Interaction, Large Displays, Human-Computer Interaction, Data Visualization, Tangible Interaction, Conceptual Metaphor Theory, Organic HDI, Casual Visualizations, Perspective Taking, Correlation, Causation, Elicitation Studies, Gesture Taxonomies

Contents

Figure Credits List

Figure 1.2   From Milka Trajkova, A'aeshah Alhakamy, Francesco Cafaro, Sanika Vedak, Rashmi Mallappa, and Sreekanth R. Kankara. Exploring Casual COVID-19 Data Visualizations on Twitter: Topics and Challenges. *Informatics*, 7(3) 35. 2020. https://doi.org/10.3390/informatics 7030035. ⓒ Multidisciplinary Digital Publishing Institute (MDPI). Used with Permission.

Figure 1.4   From http://www.inventinginteractive.com/wpcontent/uploads/2010/03 /videoplace_02.jpg. ⓒ Inventing Interactive. Used with Permission.

Figure 2.1   From Niklas Elmqvist. Embodied human-data interaction. In ACM CHI 2011 Workshop "Embodied Interaction: Theory and Practice in HCI," pp. 104–107. 2011. ⓒ Niklas Elmqvist. Used with Permission.

Figure 2.2   From Milka Trajkova, A'aeshah Alhakamy, Francesco Cafaro, Rashmi Mallappa, and Sreekanth R. Kankara. Move Your Body: Engaging Museum Visitors with Human-Data Interaction. *CHI '20: Proceedings of the 2020 CHI Conference on Human Factors in Computing Systems*, pp. 1–13. 2020. https://doi.org/10.1145/3313831.3376186. ⓒ Association for Computing Machinery Inc. Reprinted by permission.

Figure 3.3   From Alissa N Antle, Greg Corness, Saskia Bakker, Milena Droumeva, Elise van den Hoven, and Allen Bevans. Designing to Support Reasoned Imagination through Rmbodied Metaphor. In *C&C '09: Proceedings of the Seventh ACM Conference on Creativity and Cognition*, pp. 275–284. 2009. https://doi.org/10.1145/1640233.1640275. ⓒ Association for Computing Machinery Inc. Reprinted by permission.

Figure 3.6   From Christian Schönauer, Thomas Pintaric, and Hannes Kaufmann. Full Body Interaction for Serious Games in Motor Rehabilitation. In *AH '11 Proceedings of the 2nd Augmented Human International Conference*, pp. 1–8. 2011. https://doi.org/10.1145/1959826.1959830. ⓒ Association for Computing Machinery Inc. Reprinted by permission.

Figure 3.7 From David Birchfield, Thomas Ciufo, and Gary Minyard. SMALLab: A Mediated Platform for Education. In *SIGGRAPH '06: ACM SIGGRAPH 2006 Educators Program*, pp. 33–es. 2006. https://doi.org/10.1145/1179295.1179329. © Association for Computing Machinery Inc. Reprinted by permission.

Figure 3.8 From Remo Pillat, Arjun Nagendran, and Robb Lindgren. A Mixed Reality System for Teaching STEM Content using Embodied Learning and Whole-body Metaphors. In *VRCAI '12: Proceedings of the 11th ACM SIGGRAPH International Conference on Virtual-Reality Continuum and Its Applications in Industry*, pp. 295–302. 2012. https://doi.org/10.1145/2407516.2407584. © Association for Computing Machinery Inc. Reprinted by permission.

Figure 3.9 From Ayelet Segal. Do Gestural Interfaces Promote Thinking? Embodied Interaction: Congruent Gestures and Direct-touch Promote Performance in Math. Ph.D. Thesis, Columbia University, 2011. https://doi.org/10.7916/D8DR32GK. © Ayelet Segal. Used with Permission.

Figure 4.1 From John H. Falk and Lynn D. Dierking. *Learning from Museums, Second Edition.* 2018. © Rowman & Littledield. Used with Permission.

Figure 4.3 From Leslie J. Atkins, Lisanne Velez, David Goudy, and Kevin N. Dunbar. The Unintended Effects of Interactive Objets and Labels in the Science Museum. *Science Education* (93)1, pp. 161–184. 2008. https://doi.org/10.1002/sce.20291. © Wiley Periodicals, LLC. Used with Permission.

Figure 4.4 From Cherry Thian. Augmented Reality—What Reality Can We Learn From It. "Museums and the Web," 2012. https://www.museumsandtheweb.com/mw2012/papers/augmented_reality_what_reality_can_we_learn_fr © 2012 Museums and the Web LLC. Used with Permission.

Figure 7.1 From Michael D. Good, John A. Whiteside, Dennis R. Wixon, and Sandra J. Jones. Building a User-Derived Interface. In *Communications of the ACM*, (27) 10, pp. 1032–1043. 1984. https://doi.org/10.1145/358274.358284. © Association for Computing Machinery Inc. Reprinted by permission.

Figure 7.2 From Museo8bits, CC BY-SA 3.0, https://commons.wikimedia.org/w/index.php?curid=944858.

Figure 7.3   From Jacob O.Wobbrock, Meredith Ringel Morris, and Andrew D. Wilson. User-defined gestures for surface computing. In *CHI '09: Proceedings of the SIGCHI Conference on Human Factors in Computing Systems*, pp. 1083–1092. 2009. https://doi.org/10.1145/1518701 .1518866. © Association for Computing Machinery Inc. Reprinted by permission.

Figure 7.4   Ibid.

Figure 7.5   From Francesco Cafaro, Leilah Lyons, and Alissa N. Antle. Framed Guessability: Improving the Discoverability of Gestures and Body Movements for Full-Body Interaction. *In CHI '18: Proceedings of the 2018 CHI Conference on Human Factors in Computing Systems*, pp. 1–12. 2018. https://doi.org/10.1145/3173574.3174167. © Association for Computing Machinery Inc. Reprinted by permission.

Figure 7.6   Ibid.

Figure 8.2   From Jessica Roberts, Francesco Cafaro, Raymond Kang, Kristen Vogt, Leilah Lyons, and Josh Radinsky. That's Me and That's You: Museum Visitors' Perspective-Taking Around an Embodied Interaction Data Map Display. In N. Rummel, M. Kapur, M. Nathan, and S. Puntambekar (Eds.), *To See the World and a Grain of Sand: Learning across Levels of Space, Time, and Scale: CSCL 2013 Conference Proceedings, Volume 2—Short Papers, Panels, Posters, Demos, & Community Events*, pp. 343-344. 2013. https://repository.isls.org//handle/1/1901. © International Society of the Learning Sciences (ISLS). Used with Permission.

Figure 8.3   From Jessica Roberts, Leilah Lyons, Francesco Cafaro, and Rebecca Eydt. Interpreting Data from Within: Supporting Humandata Interaction in Museum Exhibits through Perspective Taking. In *IDC '14: Proceedings of the 2014 Conference on Interaction Design and Children*, pp. 7–16. 2014. https://doi.org/10.1145/2593968.2593974. © Association for Computing Machinery Inc. Reprinted by permission.

Figure 8.4   From Jessica Roberts and Leilah Lyons. The Value of Learning Talk: Applying a Novel Dialogue Scoring Method to Inform Interaction Design in an Open-Ended, Embodied Museum Exhibit. In *International Journal of Computer-Supported Collaborative Learning*, (12), pp. 343–376. 2017. http://dx.doi.org/10.1007/s11412-017-9262-x. © International Society of the Learning Sciences (ISLS). Springer Nature. Used with Permission.

Figure 8.5   Ibid.

Figure 8.7 From Jörn Hurtienne and Johann Habakuk Israel. Image Schemas and Their Metaphorical Extensions: Intuitive Patterns for Tangible Interaction. In *TEI '07: Proceedings of the 1st International Conference on Tangible and Embedded Interaction*, pp. 127–134. 2007. https ://doi.org/10.1145/1226969.1226996. © Association for Computing Machinery Inc. Reprinted by permission.

Foreword by Niklas Elmqvist

As I write this one rainy afternoon in April 2021, we have gone well past the one-year point of the global COVID-19 pandemic. It is only in writing this foreword that it strikes me what I—and, more broadly, the human race—have lost as a victim of this pandemic: physical place and the role our bodies play in it.

It also never occurred to me until now that this very book that Francesco and Jessica have written, the book you are right now holding in your hand (or, more likely, viewing through your electronic reader), is anchored in that fundamental concept of embodiment that has so suffered under the social distancing guidelines necessary for our society to conquer this pandemic.

My own college, the College of Information Studies at the University of Maryland, College Park, USA, is a former library school, and librarians and archivists are certainly some of the original practitioners of human-data interaction. After all, what is a librarian if not an information professional who is intimately familiar with the activities of seeking, processing, and interacting with vast amounts of data? However, I was aware of none of that when I published my initial paper on embodied human-data interaction at an embodied interaction workshop at the ACM CHI conference in 2011. Accordingly, my four-pager only scratched the surface of this exciting idea for combining tangible and embodied interaction with sensemaking.

What Jessica and Francesco have done in this book is to give this topic the proper treatment it deserves: to (1) study its theoretical underpinnings in perceptual, cognitive, and learning sciences; (2) draw inspiration, derive problems, and build models based on museums and other informal learning environments; and (3) present a careful, complete, and thorough review of how to design embodied human-data interaction tools for all of these settings. The book does a wonderful job of putting the personal perspective back into how humans perceive and make sense of data. It does so by leveraging a host of post-WIMP interfaces such as touch, gestures, and full-body interaction. These guidelines and findings can be a great help for practitioners, researchers, and educators alike looking to apply these ideas to real-world learning environments and museum exhibits.

At the time of writing, vaccinations are happening all over the world, and we can only hope that the COVID-19 pandemic will eventually be no more than a particularly unpleasant chapter in the history of the world. Perhaps then we will be able to go back to our museums, to our classrooms, to our playgrounds and preschools. And hopefully then, Francesco's and Jessica's work will be there to help us—and, more importantly, our children—learn and experience the world of data that surrounds us.

Niklas Elmqvist
Annapolis, MD, USA, April 2021

Acknowledgments

Our work in human-data interaction (HDI) started more than 11 years ago, when we began collaborating to build and test interactive systems to engage people with data visualizations. At that time, we were both Ph.D. students at the University of Illinois at Chicago, working under the supervision of Leilah Lyons. It is needless to say that we are deeply indebted to our common advisor, both for her invaluable mentorship and for the opportunity that she offered us to do user-engaged work at museums in Chicago and New York. It is thanks to the discussions we had with Dr. Lyons and Dr. Josh Radinsky that we originally began exploring HDI.

This book would not exist without Niklas Elmqvist, David Ebert, and Diane Cerra, who encouraged us to write it and provided continuous feedback on our chapters. Thanks for being so flexible and patient with us during this process! Special thanks also to Ben Shapiro for carefully reading an initial draft of our book and providing many pages of insightful suggestions.

Thanks to the Ph.D. students who actively collaborated on the research in HDI that we describe in this book, in particular Milka Trajkova and A'aeshah Alhakamy; to the many graduate students who contributed to the design and implementation of IDEA, including Swati Mishra, Sreekanth Kankara, Rashmi Mallappa, Sanika Vedak, and Hinal Kiri; and to the graduate students who provided feedback on a very early draft of the background chapters of this book (Ulka Patil, Sowmya Chandra, Gennie Mansi, and Abhijeet Saxena). Thanks to the many colleagues with whom—through many years—we discussed topics that ended up in this book, in particular Robb Lindgren, Alissa Antle, Michael Horn, Andy Johnson, Susan Goldman, Tom Moher, Steve Mannheimer, Erin Brady, and Davide Bolchini, as well as the Georgia Tech Vis Lab led by John Stasko and Alex Endert.

Thanks to the staff and directors at the New York Hall of Science in Queens, NY (in particular, Rebecca Eydt), at the Jane Addams Hull House Museum in Chicago, Illinois, and at Discovery Place Science in Charlotte, NC (in particular, Tifferney White, Samantha Wagner, Kellen Nixon, and Kaylan Petrie).

We also want to acknowledge that portions of the work that we describe in this book were made possible thanks to various federal grants. CoCensus was supported by the National Endowment for the Humanities Digital Startup Grant and by the National Science Foundation under INSPIRE grant no. 1248052 (Studying and Promoting Quantitative and Spatial Reasoning with Complex Visual Data across School, Museum, and WebMedia Contexts). IDEA was supported by the National Science Foundation under CHS:EAGER grant no. 1848898 (Aiding Reasoning about Correlation and Causation). Some of the exploratory research on how to use Conceptual Metaphor Theory for designing embodied technologies was supported by the National Science Foundation under FW-HTF-P grant no. 1928549 (Building Research Capacity by Technological Interventions in Support of Mixed-Ability Workplaces).

Finally, this book would not have been possible without the encouragement, patience, and help of our families—especially during the extraordinary time that we all endured with the COVID pandemic in 2020. Thanks Stella, Anthony, Sofia, George, Quentin, and Violet!

Francesco Cafaro and Jessica Roberts
May 2021

C H A P T E R   1

Introduction

Take a moment to think about how you have interacted with data today. Did you click through personal or business financial reports? Maybe you logged personal health information such as your weight, exercise, or blood pressure, or perhaps you relied on a fitness tracker to log this information for you. You may have investigated a data representation paired with an article as you scrolled through headlines in your preferred news application. Whether any or all of these specific examples match your experiences, chances are that most days you actively engage with quantified data in some way in your personal life. Though it seems natural now, it was not long ago that this level of interaction with and access to such data was unusual, except for those working with data professionally.

Beyond merely increasing frequency of data interactions, the 21st century and its technical and networking advancements have shifted public data engagements from passive consumption of data to active interactions with data. Online open data repositories allow practically anyone to download and explore large sets of data on virtually any topic. Moreover, a broad array of accessible visualization tools permits creation of professional-looking data representations without the need for professional training: a few clicks and your own uploaded dataset is visualized within minutes!

This democratization of data—providing access of information to all people, rather than limiting access to elite professionals in business and government—has been a professed aim of democratic societies since the advances in networked computing made it a possibility in the 1990s (Sawicki and Craig [1996]). While widespread data access is closer than ever—though notably not equitably distributed across racial and socioeconomic groups (Peck et al. [2019])—we cannot let the regularity of data interactions mask the inherent challenges of data interpretation, particularly when the dataset itself is unfamiliar to the user (Börner et al. [2016]). Fortunately, this expanded access to data has occurred alongside—and is inherently intertwined with—advances in personal computing technologies and infrastructure. From touchscreens to motion-sensing technologies, we have the ability to manipulate digital data far beyond the single-user windows-icons-menus-pointer (WIMP) interactions of the past. The myriad of available technologies and interfaces opens seemingly limitless possibilities for mediating human-data interactions (see Figure 1.1).

With these opportunities, though, come novel design challenges. How can we harness technologies to support meaningful human-data interactions? And, what are the affordances of off-the-desktop interactive systems for promoting data literacy?

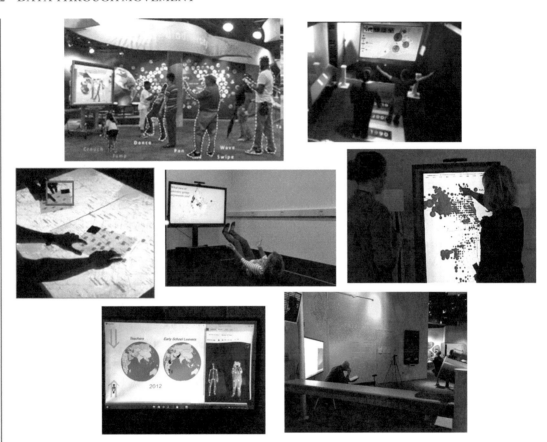

FIGURE 1.1: Human-data interactions (HDI) take many forms as sensing and input technologies allow users to control interactive data visualizations with hand gestures and body movements. This book will explore the design, functionality, and theoretical underpinnings of HDI installations and their implications for learning.

1.1   Modern Data Literacy

Two centuries ago, being able to write one's name or recite memorized text qualified a citizen as sufficiently literate. As demands on workers' skills increased throughout the 20th century, so did the requirements of literacy: first to be able to read and comprehend unfamiliar text and then to be able to make inferences and solve problems by synthesizing knowledge across multiple sources (Bransford et al. [2000]). Similarly, the expansion in recent decades of the role data plays in our daily lives has revealed the need for a similar expansion in the definition of *data literacy* (Frank et al. [2016]). The requirement to read, understand, and interpret data representations persists, but the complexity of these representations has increased exponentially, outpacing efforts to scaffold people's interpretations of them.

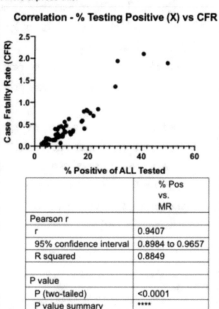

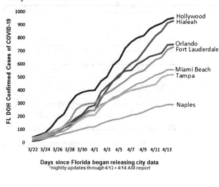

FIGURE 1.2: Examples of casual data visualizations that Twitter users spontaneously created and posted during the early stages of the COVID-19 pandemic. (*Source*: Trajkova et al. [2020b].)

Card et al. [1999] give us an oft-cited definition of data visualization as "the use of computer-supported, interactive visual representations of abstract data to amplify cognition." Data scientists will use visualization as a jumping-off point for exploratory analysis, to look for patterns and ask questions about their data. The aforementioned expansion of access to data and visualization tools has afforded such exploratory analysis for lay users as well. We saw firsthand the extent to which these handcrafted visualizations have proliferated when, during the COVID-19 pandemic, data representations created by users from all backgrounds and perspectives were shared profusely on Twitter and other social media as members of the public tried to grapple with the novel and deadly disease (see examples in Figure 1.2).

Data literacy, then, is an increasingly crucial skill set as more and more people rely on their own data analyses to inform their personal and business decisions (for example, to determine which activities are safe to do during a pandemic). Thus, empowering people to understand the data that surrounds them has become a crucial goal for interaction (IxD) and user experience (UX) designers, as it enables people to exercise agency in controlling and promoting their choices and their society's well-being.

This book explores the design of interactions with public installations that facilitate exploration and sense making of large sets of data (see Figure 1.1). We refer to this sub-field of interaction design (Rogers et al. [2011]) as "human-data interaction" (HDI). This term was used in the early 2010s in Elmqvist [2011] and Cafaro [2012] to refer to a research stream that investigates how users interact with large sets of data using novel interfaces and has since been picked up and used across multiple communities (Victorelli et al. [2019]). We draw on our work over the past decade designing, developing, and testing data engagement experiences for non-expert users primarily in museums.

We approach this topic from a unique perspective: neither author comes from an information visualization background. Francesco Cafaro was trained in computer science and currently works in human-computer interaction (HCI); Jessica Roberts came to the study of data representations from the learning sciences, a discipline that emerged in the 1990s from cognitive science, computer science, and education, and currently works in interactive computing. We began collaborating in 2010 to build and test interactive systems to engage visitors with data, and we hope with this volume to begin translating and synthesizing our multi-disciplinary insights to HDI.

1.2   Data Interactions through Embodiment

When you picture human-data interaction, what does it look like? Is it a user working alone at a computer, manipulating a spreadsheet or a 2D graph using keyboard and mouse inputs? Technological advances in recent years have allowed us to challenge this decades-old paradigm for data interactions. A growing body of work is investigating the power of large-screen displays (Andrews et al. [2010], Ball and North [2007], Liu et al. [2016], Prouzeau et al. [2016]) and multi-touch tables (Davis et al. [2013], Elmqvist [2011], Ma et al. [2012]) to create collaborative multi-user data displays. For example, Hurtienne et al. [2020] describe Move&Find, an installation in which people have to pedal on a bike to visualize "the energy required to power a search query on Google." Moreover, advances in motion-sensing devices, for example, the Microsoft Kinect and Ultraleap's Leap Motion Controller, have opened up possibilities for interacting with data not through the windows-icons-menus-pointer (WIMP) input paradigm (Hutchins et al. [1985], A. Van Dam [1997]) but instead through bodily movements and gestures.

This concept of *embodied human-data interactions* has emerged as a way to capitalize on theories of embodied cognition—the idea that the body plays a fundamental role in human cognitive processes—to promote deeper engagement with abstract data. Embodied interaction is the practical application of embodied cognition to human-computer interaction (Dourish [2001]). Hornecker [2011] contextualizes embodied interaction thusly:

Dourish highlights how embodied interaction is grounded and situated in everyday practice, constituting a direct and engaged participation in a world that we interact with. Through this engaged interaction, meaning is created, discovered, and shared. Embodied interaction is thus socially and culturally situated. But phenomenologies and ecological psychologies would argue that being situated also means being situated in a body: Your body affects your experience of the world, changing your viewpoint quite literally as well as your experience of the world in terms of what it allows you to do (Husserl's "I can's"). In this sense the physicality of our bodies is tightly linked with our experience of the physicality of our surroundings.

Chapter 3 of this book will explore the theoretical foundations of "embodiment": we will review theories from cognitive science, cognitive linguistics, and education that explain how learning happens through our physical (i.e., bodily) interaction with the world. As has been previously described, however, designing gestural interaction can be a challenging task for traditional HCI design methods (Maher and Lee [2017]). We discuss methodological implications (and approaches) for HDI in Chapter 7.

1.3   HDI in Informal Spaces

While humans interact with data across many or all contexts in our lives, studies examining data sensemaking tend to be situated in workplaces, school, or lab interaction experiences. This book will focus on informal learning and, in particular, museums. As the gathering, analysis, and visualization of large datasets become increasingly central to modern scientific activity, museums are increasingly attending to ways of presenting authentic data to visitors (Ma et al. [2019], J. Roberts and Lyons [2020]). However, as "designed spaces" for informal learning (National Research Council [2009]), museums aim to support social learning, inspire curiosity, and engage visitors with new ideas and phenomena that they can connect to their prior knowledge and understandings. This type of engagement may be difficult to support with a traditional single-user WIMP system (see Figure 1.3).

As discussed in detail in a book by Hornecker and Ciolfi [2019], interactive installations have the potential to reshape museum galleries and the visitors' experience. In particular, we believe that the unique social, interactive nature of museums makes them ideal for exploring the potential for off-the-desktop interfaces involving whole-body or full-body interaction. Museums have embraced gestural interaction: its novelty still generates buzz and excitement among their patrons and it has enormous educational potential. For example, work has shown that children better understand and remember physics concepts when they are asked to "embody" a meteor in an interactive simulation (i.e., to run through a room to control a floor-based display), rather than when they are introduced to similar concepts using a more traditional desktop simulation (Lindgren and Johnson-Glenberg [2013]). Chapter 4 will dig into the nature of museum (and other informal) learning. We will

FIGURE 1.3: This tongue-in-cheek exhibit idea from Leilah Lyons, the principal investigator on the CoCensus project described later in this book, captures the tension of translating data science into museum engagement experiences. As data infiltrates our lives beyond work and school, where, why, and how we interact with it must change too. Whole-body and gestural interactions and large, multi-user displays have the potential to bring HDI out of single-user spreadsheets and into social, collaborative spaces.

describe how museums are inherently social spaces (people typically visit museums in groups, rather than alone) and discuss opportunities for additional knowledge construction as people may bring their personal experience to the table when they talk with each other through the museum space. We will review a variety of ways multi-user technologies have been designed for museums and discuss the learning metrics by which they are evaluated.

1.4   Our Perspective on Human-Data Interaction

We want to highlight that the focus on the design of the interaction complements a much larger research endeavor that investigates how elements of the data visualizations (lines, colors, etc.) can facilitate sense making. From our perspective, visualization and interaction are two sides of the same coin: one focuses on the visual elements on the screen, the other on the actions that people can do to explore and analyze data.

While we could not endeavor in a single volume to cover the full body of work examining sensemaking around visualizations, we commit Chapter 5 to a brief overview on the use of data visualizations as learning tools in educational settings.

The use of gestural interfaces—rather than keyboard and mouse—is what distinguishes HDI from more traditional ways of designing the interaction with data visualizations. This is the reason why we use the term "embodied" to refer to the HDI work that we illustrate in this book. As noted

FIGURE 1.4: Implemented in the 1970s by computer artist Myron Krueger, Videoplace looks very similar to modern Kinect installations. The prototype allowed people in different rooms to interact with each other using gestures and body movements. (*Source*: Image retrieved online on February 3, 2021, at: http://www.inventinginteractive.com/wp-content/uploads/2010/03/videoplace_02.jpg.)

in Elmqvist [2011], data analysis and visual analytic tools are typically based on traditional windows-icons-menus-pointer (WIMP) interfaces. Indeed, these tools fail to capitalize on the ways in which the interaction itself contributes to people's meaning-making.

Gesture-based interfaces are not that new: one of the first examples, Videoplace (Figure 1.4), was designed and implemented in the 1970s. They have been, however, largely confined to the realm of research laboratories for many decades.

Chapter 2 will illustrate how our definition of HDI differs from other common uses of this term—with particular reference to Mortier et al. [2014]'s focus on how people engage with their "personal" data—and discuss strategies to engage museum visitors with HDI installations. The definitions of HDI on which this book is based roughly coincide with the availability of commercial motion tracking devices, like the Nintendo Wii and the Microsoft Kinect. Those devices were initially met with great enthusiasm by early-adopters and by the research community because of their promise to bring gestural interaction to the masses. Such installations, however, have no utility if people are not able to operate them. This is a non-trivial task for designers, because of the novelty of these installations and the lack of standardized gestures and body movements. Thus, Chapter 7 will discuss relevant gesture classification taxonomies and illustrate design strategies and methodologies to facilitate gesture design, with a particular focus on elicitation studies (Wobbrock et al. [2005], Cafaro et al. [2018]). Finally, Chapter 8 will present two use cases that exemplify how HDI can facilitate data sense making in informal learning settings.

We do not begin to claim that this book can fully cover this vast and rapidly growing area that goes under the umbrella term "human-data interaction." Instead, we present the foundational concepts that have grounded our own work in designing HDI for informal learning environments with the hopes of inspiring conversation and future work on the inherent and exciting challenges in this domain.

CHAPTER 2

Understanding Human-Data Interaction

In this chapter, we explain how our definition of human-data interaction is strongly rooted in inter-action design and embodied interaction and how, as such, it differs from other ideas of HDI, with particular regards to Mortier et al.'s [2014]. Next, we distinguish *organic HDI* as a particular class of HDI occurring in informal environments: users approach a data visualization without a distinct informational goal, and the casual nature of this interaction impacts their sense making.

2.1 HDI: A Broad, Multi-Disciplinary Research Field

In their chapter on HDI in the *Encyclopedia of Human-Computer Interaction*,[1] Mortier et al. observe that there are distinct "versions" of HDI. Similarly, Victorelli et al. [2019] conducted an extensive review of HDI literature and reported that the term "HDI" has been used to refer to a diverse range of research topics and application domains. This can be confusing for the reader of this book, because of the different expectations that the term may evoke.

Thus, we want to provide some clarity on how this book is positioned within current HDI research and to define the boundaries and scope of the embodied interaction work that we discuss. In this section, we review different definitions of HDI to help the reader with the challenging task of navigating HDI literature.

2.1.1 HDI in Data Visualization

First of all, we need to clarify that the wording "Human-Data Interaction" is not new. To the best of our knowledge, the term was used for the first time back in the late 1980s. In particular, Donoho et al. [1988] use the term to denote a process in which analysts can "transform, edit, and categorize data" by using a graphical user interface. The focus of that work was on illustrating how a dynamic graphics system (i.e., animations that show change over time) can be used to identify patterns in data. The specific use case for that work was car road-test scores from Consumer Reports.

---

1. Available online at: https://www.interaction-design.org/literature/book/the-encyclopedia-of-human-computer-interaction-2nd-ed.

Similarly, Zhu et al. [2008] use HDI to refer to hierarchical data visualization tools that allow users to analyze multi-variate data. A more recent workshop paper (Cabitza et al. [2016]) points out the need to investigate how data visualization tools influence the way in which people make sense of healthcare data.

A common thread across these works is the focus on how interactive data visualizations can help people with making sense of complex data and with finding patterns in data.

2.1.2   Early HDI Work on Databases and Information Systems

Early work on HDI, however, mostly focused on the process of interacting with databases. For example, Kennedy et al. [1996] use the term "human-data interaction" when describing an interface that allows users to visually create and update database queries. They explain that "the visualisation [of the data elements in the database] and the interaction together constitute the interface between the user and the database"—an idea that is very similar to our focus on the interaction aspects of HDI.

In Hill and Aspinall [2000], "human-data interaction" is used to denote the need to "link database entities with time attributes to an animation or video presentation, in order to satisfy the user's information needs."

In this case, the common thread (across this line of early HDI work) is in the use of graphical user interfaces to help with navigating complex database structures and finding data.

2.1.3   HDI and Personal Data

More recently, Mortier et al. [2013] define HDI as a multi-disciplinary line of research that investigates how people can engage with personal data "whether as users of online systems or as subjects of data collection." In later work, Mortier et al. [2014] further highlight the HDI focus on "empowering us to become aware of the fact and implications of collection of our personal data."

In summary, Mortier's line of work investigates how personal data is collected and shared, the ethical implications surrounding the collection and use of data, and issues related with privacy and consent. For example, Sailaja et al. [2021] describes the design of a web-based interface that allows users to manage permissions across multiple media channels. Similarly, the work in Mashhadi et al. [2014] poses the problem of preserving ownership of the data that is now collected by a multitude of IoT devices.

Although this line of research is foundational for HDI in a broad sense, it is very much delimited by its focus on personal data. Thus, it does not overlap much with the embodied interaction approach to HDI that we follow in this book.

FIGURE 2.1: Transparent plastic sheets are used to reveal different layers of a map-based visualization on a tabletop display. (*Source*: Elmqvist [2011].)

FIGURE 2.2: Museum visitors interact with data visualization using gestures and body movements. (*Source*: Trajkova [2020a].)

2.1.4   HDI and Embodied Interaction

In the early 2010s, Elmqvist [2011] and Cafaro [2012] introduced a different perspective on HDI. These works investigate how users *interact* with large sets of data that are visualized on novel interactive displays (e.g., gesture-controlled large displays and tabletops).

For example, Figure 2.1 shows plastic sheets that are used as *tangible* lenses to reveal different layers of an interactive map. Figure 2.2, instead, displays museum visitors interacting with a globe-based data visualization using gestures and body movements (that are recognized thanks to a Microsoft Kinect tracking camera).

As noted in Elmqvist [2011], data analysis and visual analytic tools (Wong and Thomas [2004]) are typically based on traditional windows-icons-menus-pointer (WIMP) interfaces.

Indeed, these tools fail to capitalize on the ways in which the "interaction" itself contributes to people's meaning-making when they are exposed to large sets of data.

The reader should notice that this research perspective is deeply rooted in embodied interaction, embodied cognition, and embodied learning: the idea here is that the gestures and body movements that we perform or the tangible object that we manipulate alter the mental patterns that are *preconceptually* activated in our brain. Ultimately, the focus of this embodied perspective to HDI is on investigating how the *interaction* with a data visualization can mediate people's exploration and understanding of the data on display. We will explore the conceptual roots of embodied cognition and embodied interaction in Chapter 3.

2.1.5   Common Threads in HDI

In conclusion, "human-data interaction" is an umbrella term. This is in the broad definition provided by Victorelli et al. [2019]:

> [HDI is a research] area addressing human manipulation, analysis, and sensemaking of large, unstructured, and complex datasets, understanding their meaning as well as considering stakeholders and the data life cycle phases.

Despite the divergent research foci, methodologies, and approaches, there are at least three common threads in HDI. First, the focus on data and on people's interaction with data. Second, the multi-disciplinary nature of HDI work. Third, the fact that the user is at the center of the design process (Locoro [2018]), regardless of whether the design involves windows-icons-menus-pointer (WIMP) interfaces to query databases, information systems that empower users with their personal data, or interactive visualizations that facilitate data sense making.

2.2   Motivation in HDI

What motivates someone to interact with a dataset? Analysts and data scientists, for whom many visualization platforms are designed, explore data to uncover insights, whether beginning with a targeted question or engaging in open-ended exploration to find a pattern worth investigating. It has long been acknowledged that lay users, i.e., those not trained in data science or information visualization, are also exploring visualized data, leading to a need for collaborative platforms for non-technical users (Heer et al. [2008]). The term "casual infovis" has been used to describe artistic, social, and ambient information visualizations that are "a part of, but different from, more traditional infovis systems and techniques" (Pousman et al. [2007]). When conducting an observation study with 22 non-experts, for example, Sprague and Tory [2012] found that combinations of extrinsic (e.g., "avoiding boredom") and intrinsic motivations (e.g., "curiosity") promoted the exploration of casual data visualizations, and they observed that creating a sense of "personal relevance" is critical to enable longer and repeated interactions.

Particularly prominent in recent years is the increase of complex interactive data vis in news and information sources for the general public. As an example, when the respiratory infection COVID-19 (a.k.a. Sars-CoV-2 or "the novel coronavirus") began its spread across the globe in late 2019 through 2020, visualized datasets soon followed. Maps plotted outbreak hotspots, column charts broke down mortality rates by age bracket and other demographics, and line graphs plotted new cases (which became even more frightening when people realized these graphs were logarithmic, not linear). Epidemiologists and health officials set to work to use these data to find answers. An anxious public, meanwhile, began poring over these often dynamic and interactive web-based visualizations to make sense of this novel disease for themselves. They asked, "Where am I in this dataset?" and "What's my risk?" Some relied exclusively on premade representations, and others downloaded data to create their own visualizations for exploration (Trajkova et al. [2020b]).

Though most members of the public were not exploring the data in order to solve the larger global pandemic, their motivations for exploration were targeted, much like traditional vis. That is, they were asking questions of data and the answers were important to them. We might think of these as *motivated* human-data interactions. Whether the motivation came from personal interest, as in this example, or from an instructor or supervisor in a formal school or work setting is irrelevant; what matters is the existence of an informational goal.

2.2.1   Organic HDI

Contrast the above example with how these same people might approach a map of bird migrations. Everyone has seen birds, and most people can identify at least a few species. Some birds move more than others, causing visible variation in patterns. When these migration data are mapped onto interactive displays, such as was done by the Cornell Lab of Ornithology in their eBird platform (Figure 2.3), they can illuminate fascinating large-scale patterns that are otherwise invisible to a lay observer.

These eye-catching maps are similar to what would be encountered in a museum exhibit. In museums, visitors will often approach an exhibit in "browse mode" (Schauble et al. [2002])—not because of a specific interest in or need from the particular exhibit, but because it is available for exploration. An exhibit interaction typically begins, therefore, without a set informational goal. It develops organically and persists as long as it remains interesting to the visitor. Unlike professionals'

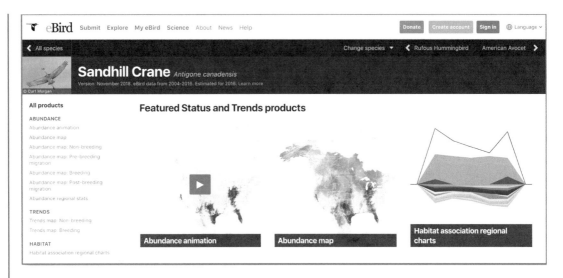

FIGURE 2.3: The Cornell Lab of Ornithology's Status and Trends products are based on citizen-generated bird sighting data. (*Source*: https://ebird.org.)

interactions with visualizations, visitors have no mandate to find insights in their data. Unlike artistic and infographic casual infovis interactions, the visitors cannot be assumed to have preexisting interest or motivations in the data. And unlike data encounters in formal educational environments, visitors receive little or no feedback from a more expert instructor about whether any insights they do uncover are correct.

We argue these differences add up to significant deviations from much prior work in HDI. Though museum researchers have begun addressing questions of how lay visitors interpret graphical data displays (Börner et al. [2016]; Ma et al. [2020]), to our knowledge there is not an existing term in the visualization literature capturing this unique user scenario. We dub these interactions *organic human-data interactions* and posit they are different learning interactions than the traditional and casual motivated interactions described above.

Organic HDI is particularly suited for embodied interaction design, particularly in museums. The physical embodiment of data may have affordances for supported sensemaking, even with unfamiliar and complex datasets (J. Roberts and Lyons [2020]). As we will show in the upcoming chapters, designing interactive data visualizations that support organic HDI is, however, challenging, beginning with the intricacies of gestures in human communication.

2.3   Summary: Embodied, Organic HDI

In this chapter, we reviewed definitions of "human-data interaction" and explained the book's focus on the design of embodied experiences with interactive data visualizations. We then introduced the concept of organic HDI and described how this is typically the type of casual data exploration that occurs in informal learning settings.

We mentioned "embodied interaction" and "informal learning" many times. In the next chapters, we will take a step back and provide an overview of relevant background and literature on these topics.

CHAPTER 3

Theoretical Foundations
Embodiment

Human-data interaction is grounded on sociocultural theories of cognition and learning. This chapter provides an overview on the theoretical foundations of embodiment—which we believe are essential to properly understand, design, and evaluate HDI installations. It is organized in three major sections. First, we review theories of embodied cognition that explain how learning happens through our physical interaction with the world, and we discuss Lakoff and Johnson's Conceptual Metaphor Theory (CMT). Second, we compare definitions of "embodied interaction." Third, we discuss theories of learning in informal settings, emphasizing learning as situated in a social, cultural, and historical context and mediated through interactions within that context.

3.1   Embodied Cognition

Embodied cognition is a theory in contemporary cognitive science positing the overarching idea that our body plays a fundamental role in our cognitive process: we make sense of what is around us because of our physical interaction with the world. The notion of "embodiment" is central to embodied cognition: our body constraints the concepts that we can learn (up to the point that if we had a different body, we would understand the world in a different manner), and our discoveries happen thanks to the interaction between our body and the surrounding environment (L. Shapiro [2010]).

Philosophically, embodied cognition is rooted in the phenomenology work of Husserl [2001] and Merleau-Ponty [1996]. Etymologically, the word "phenomenology" combines two Greek words, the study of ("-logy") things that appear ("phenomenon") (see Zalta et al. [1995]). Thus, phenomenology is the study of phenomena, which are seen as things that appear to us in our conscious experience. "Conscious experience" is central in phenomenology and puts the focus on phenomena that we live through, experience, or perform (Moran [2002]). Phenomenology and, in particular, Merleau-Ponty's work, criticizes the idea that the body is a mere instrument that gets orders from a transcendent mind. Rather, our body plays a fundamental role in our cognitive process (M. Wilson [2002]).

Experimentally, embodied cognition is based on a multi-disciplinary array of scientific observations in the fields of neuroscience and linguistics. For example, McNeill [1992] reports that gesturing while speaking facilitates communication. In a study using magnetic resonance, Harpaintner et al. [2020] discovered that people's sensory-motor systems are activated when they are presented with motor (e.g., "fitness") and visual (e.g., "beauty") abstract words, in the same way they were activated when the same people were asked to move their hand or look at pictures, respectively. Similarly, Dreyer et al. [2015] reports that our left sensory-motor cortex is activated when we process abstract concepts such as "love" and that lesions on the hand motor cortex may impair patients' ability to process concrete tool nouns. For an extensive introduction to embodied cognition, see A. Wilson and Foglia [2011].[1]

3.2   Lakoff and Johnson's Conceptual Metaphor Theory

In 1980, Lakoff and Johnson performed an extensive exploration of metaphors and introduced Conceptual Metaphor Theory (CMT). This theory is particularly relevant for HDI because, as we will see throughout this book, it can be used to inform and assess the design of interactive installations. Lakoff and Johnson's idea [2008] is that metaphors[2] are not simply "a device of poetic imagination"; rather, they play a fundamental role in how we understand the world and ourselves.

3.2.1   Conceptual Metaphors

According to CMT the *conceptual metaphors* that we use in our language reveal our thinking process, and in particular how we understand a concept or domain (the destination domain) in terms of another (the source domain).

One example of conceptual metaphor is ARGUMENT IS A WAR[3] (see Figure 3.1). We can understand sentences such as "your claims are indefensible" and "he shot down all my arguments"— which do not literally make sense—because they are grounded on this conceptual metaphor, and we share a common understanding of this conceptual metaphor. In other words, through this conceptual metaphor, we reason about a concept ("argument") using the same mental patterns through which we conceptualize another ("war").

It all happens extremely quickly in our mind, i.e., *preconceptually*, a term that means that this metaphor "structures the actions that we do when arguing" (Lakoff and Johnson [2008]), without our conscious realization that we are using a metaphor to decide what to do.

One important characteristic of conceptual metaphors is their *polysemy*: we may reason about the same concept using a variety of very different metaphors. For example, ARGUMENT IS A WAR is not the only way in which we talk about arguments. Alternative conceptual metaphors include ARGUMENT IS A JOURNEY (e.g., we "arrive" at a conclusion, "get to" the point) and ARGUMENT IS A

---

1. Available online at: https://plato.stanford.edu/entries/embodied-cognition/#WhaEmbCog.
2. In the *Oxford Dictionary*, a metaphor is defined as "a figure of speech in which a word or phrase [in a source domain] is applied to an object or action [in a destination domain] to which it is not literally applicable." One example is "she is a shining star."
3. Conceptual metaphors and schemata are typically capitalized in the literature.

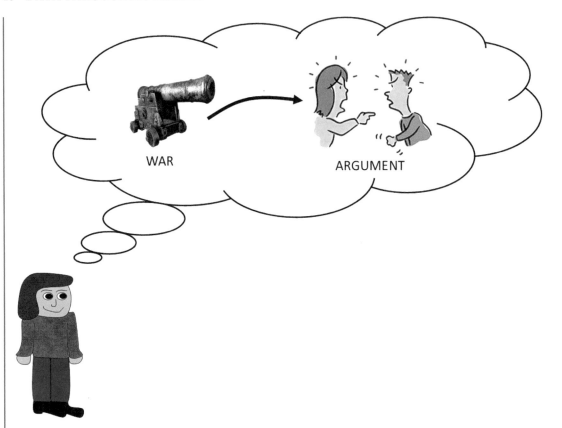

FIGURE 3.1: ARGUMENT IS A WAR is a conceptual metaphor that guides our reasoning by projecting concepts that belong to WAR (source domain) into a very different domain ARGUMENT (destination domain).

BUILDING (e.g., an argument could "collapse" or be well "constructed"). Similarly, we can use the same source domain to understand very different concepts. For example, we use WAR to understand how ARGUMENT works, but we also use it in the metaphor LOVE IS A WAR (Figure 3.2; e.g., we can try to "conquer" somebody; see Lakoff and Johnson [2008]).

3.2.2 Conceptual Primitives: Embodied Schemata

What are conceptual metaphors grounded on? In later work, M. Johnson [2013] introduces "embodied schemata"[4]. The idea is that we understand the world using a relatively small set of basic mental

---

4. "Schemata" is the plural of "schema."

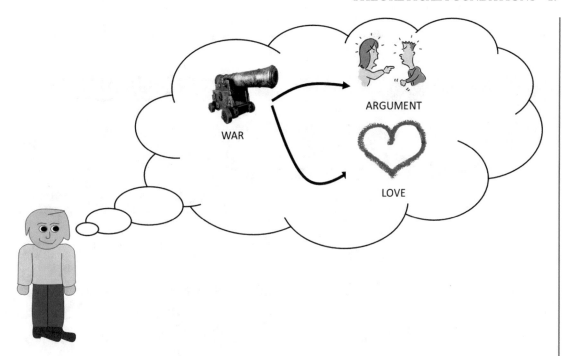

FIGURE 3.2: Polisemy of conceptual metaphors. The same source domain (WAR) can be used to interpret very different concepts (in this example, both ARGUMENT and LOVE).

patterns ("embodied schemata") that we learn at a very young age because of how we repeatedly interact with the world around us using our body.[5]

Embodied schemata arise from different types of bodily experiences, such as our need for orientation in SPACE, and the experience of an external FORCE. They emerge as "a recurrent pattern, shape and regularity in our daily body experience" (M. Johnson [2013]). For example, a baby learns the idea of BALANCE through a process of trying to stand, falling on the floor, then trying to stand over and over again, until she understands how to properly distribute her body mass. Because these mental patterns arise from very basic experiences, they are typically consistent across languages and cultures.

Similarly to conceptual metaphors, embodied schemata are identified from linguistic analysis. There is not a definitive list of embodied schemata; rather, researchers keep adding to the list and debate on what should be included. Hurtienne and Israel [2007] is perhaps to date one of the most comprehensive attempts to catalog embodied schemata in the field of Human-computer interaction and presents the list of embodied schemata that we report in Table 3.1.

---

5. The reader should notice how this notion is strongly rooted in embodied cognition.

TABLE 3.1: Catalog of Embodied Schemata

Category of Body Experience	Embodied Schemata
Space	UP-DOWN, LEFT-RIGHT, NEAR-FAR, FRONT-BACK, CENTER-PERIPHERY, STRAIGHT-CURVED, CONTACT, PATH, SCALE, LOCATION
Containment	CONTAINER, IN-OUT, CONTENT, FULL-EMPTY
Multiplicity	MERGING, COLLECTION, SPLITTING, PART-WHOLE, COUNT-MASS, LINKAGE
Force	DIVERSION, COUNTERFORCE, RESTRAINT REMOVAL, RESISTANCE, ATTRACTION, COMPULSION, BLOCKAGE, BALANCE, MOMENTUM, ENABLEMENT
Attribute	HEAVY-LIGHT, DARK-BRIGHT, BIG-SMALL, WARM-COLD, STRONG-WEAK
Process	SUPERIMPOSITION, ITERATION, CYCLE
Surface	IDENTITY FACE, MATCHING
Basic	SUBSTANCE, OBJECT

Source: Hurtienne and Israel [2007].

Importantly, embodied schemata are special concepts because they work as *conceptual primitives*: they are the foundation of our conceptual system, and conceptual metaphors are grounded on them. For example, later in life, we learn how to apply the idea of BALANCE to different, abstract domains, such as social justice and architecture. This process is called "metaphorical projection" and is at the basis of conceptual metaphors. An illustration of this process is provided in Antle et al. [2009]; see Figure 3.3.

3.2.2.1   Superimposition

Combinations of embodied schemata may produce other embodied schemata through a process called "superimposition." For example, FAR/NEAR can be superimposed to CONTAINER, generating the CENTER/PERIPHERY schema.

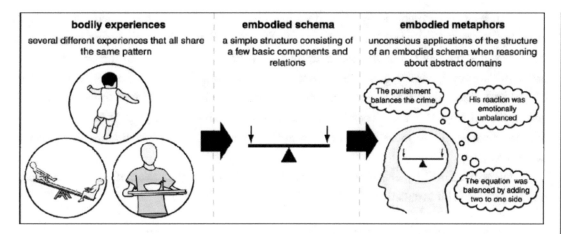

FIGURE 3.3: How repeated bodily experiences generate an embodied schema that can then be metaphorically projected to a plethora of abstract domains. (*Source*: Antle et al. [2009].)

Because embodied schemata arise when we receptively experience something with our body, the process of superimposition can only occur when we frequently experience a correlation among those schemata (M. Johnson [2013]).

3.2.2.2 Polisemy
Similarly to conceptual metaphors, embodied schemata are "polysemic": the same embodied schema may support multiple metaphorical projections into different destination domains. For instance, PATH can support metaphorical projections such as PURPOSES ARE DESTINATIONS (M. Johnson [2013]), and it can also be used in mathematics to reason about numbers as if they were disposed on a line (Lakoff and Núñez [2000]).

3.2.3 Hierarchy of Mental Patterns: Embodied Schemata, Conceptual Metaphors, and Frames
Frames are another type of mental pattern that structures our thinking. According to Fillmore et al. [2006], words in every language have the power to evoke complex scenarios. For instance, the word "restaurant" brings to mind being seated, seeing a waiter, being served food, etc. (Johnston [2002]).

Lakoff [2009] interprets frames as part of a hierarchy of mental patterns that represent actual networks of neurons that are connected with each other to form "brain circuits." Embodied schemata (the most basic mental structures) can be combined via superimposition and projected metaphorically. In turn, schemata and metaphors can be connected together in our brain in more complex networks of "brain circuits," such as frames.

Frames can be combined to form additional frames. In a keynote talk in Sarajevo,[6] Lakoff explains that the RESTAURANT frame is a connected combination of three more basic frames: (1) a FOOD SERVICE frame; (2) a BUSINESS frame; and (3) a HOST/GUEST frame. In turn, the BUSINESS frame (with a buyer, a seller, goods, and money) can be decomposed into a number of more basic mental patterns. A person is exchanging goods for money, and an EXCHANGE frame is a mutual giving, and the GIVING frame is a combination of a force embodied schema, COMPULSION (applying a force on something to move it from one location to another), and a spatial embodied schema, LOCATION.

We will see in Chapter 7 how frames may be used to inform the design of gestures and body movements for HDI installations (Cafaro et al. [2018]).

3.2.4 Effect of Polisemy: Different Mental Models Lead to Different Ways of Reasoning

As observed in M. Johnson [2013], the specific metaphors that we (preconceptually) use constrain and alter our reasoning. For example, Gentner and Gentner [2014] report that people typically reason about electricity using either a WATER-FLOW or a MOVING CROWD model. In the first case, electric current is understood as water flowing through a pipe; in the latter case, as the movement of individuals through passageways and small gates. In general, people reasoning according to the WATER-FLOW model perform better on battery problems, while those using the MOVING CROWD metaphor excel on resistor problems. This is yet another example of how conceptual metaphors are not mere linguistic devices; rather, they structure (and, potentially, alter) our reasoning.

We will discuss in Chapter 8 how this phenomenon has implications on how people make sense of data visualizations.

3.3 Embodied Interaction

As observed in Rogers [2012], much of the early human-computer interaction (HCI) borrowed methods, theories, and approaches from other disciplines. Not surprisingly, then, phenomenology (philosophy) and embodied cognition (cognitive science) are not only at the basis of CMT, but also embodied interaction, an influential theory in HCI that has inspired many works and conferences since its introduction in the early 2000s.

3.3.1 Dourish's Definition of Embodied Interaction

In 2001, Dourish published a seminal book, *Where the Action Is*, that is considered the foundation of embodied interaction. Dourish defines embodied interaction by first clarifying the meaning of

---

6. George Lakoff explains the decomposition of the RESTAURANT frame into embodied schemata in his keynote at the 4th International Conference on Foreign Language Teaching and Applied Linguistics, May 2014, Sarajevo. Retrieved online at: http://georgelakoff.com/videos/.

FIGURE 3.4: Two cubes used to control the zoom level of a map of MIT in MetaDesk. (*Source*: Ishii and Ullmer [1997].)

"embodiment": it is the way in which we make the world meaningful, i.e., by engaging and interacting with it. This definition has clear theoretical roots in the phenomenological idea of "conscious experiences," as well as in embodied cognition: according to Dourish, we construct meaning through our embodied (i.e., physical) interaction with the world. Embodied interaction is thus "the creation, manipulation, and sharing of meaning through engaged interaction with artifacts" (Dourish [2001]).

Interestingly, Dourish's definition implies a strong distinction between embodied interaction and virtual reality: virtual reality tries to bring the user into a virtual world (by creating, for example, virtual representations of the user's hands), while embodied interaction attempts to bring the virtual world into the physical world of the user (e.g., by augmenting physical objects with LED displays). This aspect of embodied interaction very much echoes Mark Weiser's [1991] idea of ubiqutious computing.

Despite its conceptual novelty, the notion of embodied interaction was based on existing technological applications. In particular, social computing (Erickson et al. [1999]) and tangible user interfaces (Ishii and Ullmer [1997]) are central for Dourish's definition of embodied interaction. In general, social computing refers to computational systems that allow people to interact with each other. Examples include customers' reviews on websites like Amazon or Tripadvisor, user-generated content such as Wikipedia, and social networks like Facebook and LinkedIn. Tangible user interfaces are an idea originally introduced by Ishii and Ullmer: "graspable" objects (such as cubes and magnifying lenses; see Figure 3.4) are used to navigate digital information. Interestingly for HDI, tangible user interfaces have later been used for navigating data visualizations. For example, Jackson et al. [2013] used a tangible user interface (a cylindrical shape prop) to control an interactive visualization of thin fiber structures.

3.3.2   The Role of the Body

As Dourish explains in a later work [2013], his 2001 book was an attempt to connect those two novel (at that time) trends in computing to the existing methodologies of the Computer Supported Collaborative Work community. As a consequence, his definition of embodied interaction did not take much of the user's *body* into account (Dourish [2013]).

Later work by Hornecker (e.g., Hornecker [2011]) instead started the process of highlighting the focus on the user's body. Hornecker posits that "tangible and embodied interaction provides a broad umbrella description for a research field united through an interest in the role of physicality. There is the physicality of our own bodies, the materiality of objects, the physical world in general, and the physicality of space." In particular, Hornecker argues that "movement and perception are tightly coupled" and explains that, for example, the way in which we experience different rooms in a building with our body can "communicate a specific atmosphere" or can "invite a particular kind of usage" (Hornecker [2011]).

This trend (increasing the centrality of the user's body in embodied interaction) is common between two slightly different approaches in interaction design: Full-body interaction and Whole-body interaction.

3.3.2.1   Full-Body Interaction

"Full-body interaction" is a term used in Hornecker and Buur [2006] to refer to the gestures and body movements that people may perform to control interactive systems. Basically, full-body interaction is a perspective on embodied interaction that mainly focuses on proving input to interactive systems.

For example, in the context of museum installations and large displays, the term "Full-body interaction" has been used to denote an "input method" in which people interact with computer systems using hand gestures and body movements (Cafaro et al. [2013]). An example is shown in Figure 3.5, which depicts a prototype museum installation that allows visitors to explore data from the U.S. census.

More broadly, full-body interaction has also been used to refer to applications that receive input from camera tracking devices and bodily sensors. Figure 3.6 illustrates an example in the context of physical rehabilitation (Schönauer et al. [2011]).

3.3.2.2   Whole-Body Interaction

Whole-body interaction is another perspective on embodied interaction that puts the focus on the design of multi-sensory experiences (England [2011]). In contrast with traditional visual interfaces, Whole-body interaction is defined by England et al. [2009] as "the integrated capture and processing of human signals from physical, physiological, cognitive and emotional sources to generate feedback to those sources for interaction in a digital environment." According to England [2011], interaction

FIGURE 3.5: CoCensus: Two users interact with a map-based data visualization using gestures and body movements that are recognized via a Microsoft Kinect tracking camera.

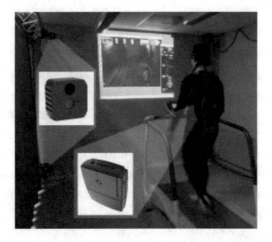

FIGURE 3.6: Mocap: A prototype system that integrates tracking cameras and bio-signals for rehabilitation. (*Source*: Schönauer et al. [2011].)

design should consider the psychical, psychological, cognitive, and emotional states of the user and use all information in an "integrated" way to both provide input to interactive systems and to structure the way in which users receive an "integrated" feedback. As a consequence, Whole-body interaction is a multi-disciplinary approach to interaction design that advocates for including research from sports, clinical settings, cognitive psychology, and arts in the design process (England et al. [2009]).

3.4   Embodiment and Learning

Though embodied cognition theory only gained traction around the turn of the 21st century, the relationships between bodies, minds, and movements have long been of interest in the study of learning. Much of the remainder of this book will explore how embodiment and embodied inter-actions can be leveraged in HDI installations, specifically to support learning and meaning-making when we interact with data visualizations. That is, how can the preconceptual mental patterns, con-ceptual metaphors, and embodied schemata described above be employed in the design of learning environments to augment learning gains and improve transfer and retention? To provide context for the design work laid out in upcoming chapters, we first will review key theories in psychology and education that have influenced related work. (For a more extensive investigation, see Greeno et al. [1996].)

3.4.1   Actions as Indicators: The Behaviorist Perspective

How do we know learning has occurred? To Behaviorists, followers of a prominent psychological movement in the mid-19th century, physical, observable behaviors were the only reliable measures of learning. Behaviorist scholars such as John B. Watson rejected the idea that psychology has to deal with consciousness, which he saw as a "neither definable nor usable concept" (Watson [2017]). The behaviorist[7] viewpoint posits that the source of human (and even animal) behavior is in the environment, not in the mind of the individual, and we learn because of environmental stimuli and conditioning. That is, manufactured stimuli—including reinforcements and punishments—provoke a reaction in the subject.

B. F. Skinner applied this theory to learning through investigations of operant conditioning, the process of applying rewards or punishments in order to continue or cease desirable and undesir-able behaviors. If a child (or a pigeon, for that matter) is rewarded for performing a certain behavior and punished for not performing the behavior, eventually the subject will learn which behavior to perform, and that performance provides a clear and measurable indication of the learning.

While the unambiguous performance indicators for assessment in the Behaviorist perspective are undoubtedly appealing, most educational researchers are unwilling to discount the role of con-sciousness and cognition in learning. Subsequent researchers and theorists in education remained

---

7. Skinner's radical behaviorism, sometimes called "big B" Behaviorism, eventually gave way to "little b" behavior-ism, a more moderate viewpoint that acknowledged some cognitive processes but still relied heavily on observable behaviors for measuring learning.

interested in behaviors, though not only as indicators of learning: more importantly, these behaviors might be integral to the learning process itself. That is, in alignment with the early work on the relationship between movement, metaphors, schemata, and cognition described earlier, physical actions not only demonstrate that learning has occurred but also support and affect how people learn. These perspectives are discussed next.

3.4.2   Constructing Knowledge

Psychologist Jean Piaget promoted the theory of constructivism, asserting that not only is there much more to learning than what is observable through behaviors, but human cognition grows through multiple stages. Children are not merely incomplete adults with empty cognitive spaces waiting to be filled. Instead, humans construct knowledge based on preexisting ideas and mental models, and even infants and small children bring these existing ideas as they attempt to make sense of new information.

The constructivist viewpoint posits that when we encounter new knowledge, we assimilate that knowledge or accommodate our knowledge structures in order to fit this new information with the old. Rarely do humans immediately adopt new ideas whole cloth and replace old ideas, even if the new idea is better. Instead, we try to fit it in and make sense of it in relation to what we already know. We test it against existing knowledge.

From a practical standpoint, the constructivist theory of learning problematizes the age-old education model of an instructor standing in front of a classroom, lecturing profoundly to an entranced audience of eager yet inert students. This *transmission model* of learning asks the learners to passively absorb information. Constructivism, on the other hand, asserts that interactivity—including physical movement—is integral to the process of testing out how new information interacts with previously held mental models. In short, we learn by doing.

These ideas were extended by Seymour Papert in his theory of *constructionism*. Papert, a mathematician who spent five years at Piaget's International Centre for Genetic Epistemology in the 1960s before moving to the MIT Media Lab, worked for nearly four decades focusing heavily on supporting learning in math. Constructionism carries the same basic tenets from Piaget—assimilation, active learning, and construction of knowledge—but emphasizes two additional components. First is the role of passion and engagement in the learning process. If a learner has a passion and develops a love and excitement for learning, it will keep her going as a learner even when mistakes are made. Importantly, asserts Papert, mistakes should be embraced as opportunities to learn and to debug, refresh, and move forward.

Papert's constructionism also places a greater focus on "body knowledge." Papert attended to not just what people are thinking but how the learning relates to our bodies and to physical objects. He believed in the pedagogical importance of "objects to think with" (Papert [2020]). Papert is perhaps best known for his contributions to the LOGO programming environment, in which learners create vectors by controlling a "turtle," iconically symbolized by a small green triangle on the screen. The turtle responds to a typed command according to its own orientation, i.e., the command LEFT would turn the turtle to its own left regardless of screen orientation. Learners thus were required to take the perspective of the turtle, embodying the view of the cursor in order to navigate. Facilitating perspective taking, as will be discussed later, is one affordance of embodied interaction design that has shown promise in fostering data interpretation (J. Roberts and Lyons [2020], Roberts et al. [2014]).

Papert distinguished between constructionism and constructivism, saying,

> Constructionism. . . shares constructivism's view of learning as "building knowledge structures" through progressive internalization of actions. . . It then adds the idea that this happens especially felicitously in a context where the learner is consciously engaged in constructing a public entity, whether it's a sand castle on the beach or a theory of the universe. (Papert and Harel [1991])

Despite this reference to external entities, constructionism—like constructivism—often forefronts a learner-centered perspective, in which the learner's own experiences, interests, skills, attitudes, and beliefs are central to the learning experience (Bransford et al. [2000]). A "public entity" for externalizing knowledge, however, hearkens to the sociocultural perspective on learning, discussed next.

3.4.3   Learning Together: A Sociocultural Perspective

The sociocultural perspective on learning views mental functioning as being situated within a cultural, historical, and institutional context, a stance that assumes that "analytic efforts that seek to account for human action by focusing on the individual agent are severely limited" (Wertsch [1998]). The perspective follows the view of Vygotsky [1978], who saw learning as occurring through dialogues, questioning, and negotiating meaning. Vygotsky introduced the concept of the Zone of Proximal Development (ZPD) to characterize the gulf where learning may take place: what a learner cannot yet do alone but can do with help. While Papert's LOGO environment was generally intended to be explored individually and wasn't intended to require curriculum or teacher support (though in practicality both were often provided to students), sociocultural environments intrinsically rely on interpersonal communication to facilitate learning.

Ash [2003] defines sociocultural settings as: "A social group or ensemble is engaged in an activity; this activity is collaborative and is informed by the individuals who comprise it and yet the activity reciprocally informs the individuals/group; the social activity is mediated by tools, signs, people, symbols, language, and actions." Research taking a sociocultural perspective on learning focuses on "meaning-making in the broad sense, which emphasizes social interaction and cultural symbols and tools as crucibles for appropriating and adapting forms of knowledge, values, and expression" (Schauble et al. [2002]).

Embodied interaction design is well aligned with the sociocultural perspective. A learner's movements are deeply embedded in the context—the environment, the other learners, and the available tools (e.g., the data visualization). Danish et al. [2020] proposed the Learning in Embodied Activity Framework (LEAF) to "synthesize across individual and sociocultural theories of learning to provide a more robust account of how the body plays a role in collaborative learning." The creation of this framework speaks to the growing body of work in the learning sciences and learning technologies exploring embodied design in social learning settings, reviewed next.

3.4.4   Embodied Learning Environments

As we mentioned at the beginning of this chapter, one experimental result at the basis of embodied cognition is the fact that words can activate specific areas of our sensory-motor cortex (e.g., Harpaintner et al. [2020]). This finding shows a connection between the actions that we do with our body and the way in which we store abstract concepts in our brain. Thus, according to Lindgren and Johnson-Glenberg [2013], if we are able to ask people to perform gestures and body movements that activate the specific areas of our sensory-motor system, we can produce "stronger and more stable memory traces and knowledge representations." This idea is at the foundation of embodied learning, a strand of work within the learning sciences that investigates how embodied cognition principles, combined with embodied interaction designs, can be used to facilitate learning.

Several notable examples of embodied learning environments exist for formal (i.e., school) and informal settings. SMALLab (Birchfield et al. [2006]) is an environment for creating room-sized embodied learning experiences. For example, an early implementation of SMALLab (illustrated in Figure 3.7) allowed students to create sounds by moving a ball in the tracking space. A more recent version of this platform is currently commercialized by SMALLab Learning[8] and used in both informal learning settings, such as the Children's Museum in Tampa, FL,[9] and classrooms (e.g., Johnson-Glenberg et al. [2009] describe an application for geology studies in high schools).

The Learning Physics through Play curriculum uses physical movements to help students learn physics concepts. For example, when children envision themselves as a ball moving across a surface, they can reason about the velocity and forces of friction of the ball in a different way than they would by just watching a ball. They are able to transform their physical bodies into "components in the microworld that structure students' inferences" (Enyedy et al. [2013]).

---

8. See: https://www.smallablearning.com/.
9. See: https://glazermuseum.org/exhibits/smallab.

FIGURE 3.7: An earlier implementation of SMALLab allowed students to generate and control sounds by moving a ball in the interaction space. (*Source*: Birchfield et al. [2006].)

Also in physics education, Danish et al. [2015] describe the STEP curriculum for teaching young students about microscopic particle movements in states of matter. Students' physical movements are mapped to dots on a large shared visualization that represent particles of matter. As students move slowly, they are acting like solid matter particles, but they can change "their" particles to gas with fast physical movements, indicated by their dots turning red. Later, Keifert et al. [2020] describe these students recreating these movements when explaining concepts in delayed interviews with the research team, sometimes even directing the researchers how to move their bodies to act like particles at different matter states. These physical re-enactments enforce the role bodily movement played in cementing the concepts for the students (Keifert et al. [2020]). This complements an increasing body of research demonstrating that hand gestures play an important role in scientific reasoning in classrooms; for example, Singer et al. [2008] found that use of gestures in a sixth-grade science class demonstrated advancement in the group's understanding of plate tectonics.

Meteor (Lindgren and Moshell [2011], Pillat et al. [2012]) is a mixed-reality (Milgram and Kishino [1994]) environment that middle school children can use to embody a meteor (see Figure 3.8). Studies of this novel museum exhibit suggest that learners utilizing a full-body simulation of kinematics were less likely to focus on "surface features" of the simulation compared to those using a desktop version of the same simulation (Lindgren and Moshell [2011]), and when the underlying concepts to be learned were logically mapped to the physical movements, learners were able to retain generative physics knowledge at a one-week follow-up better after a high-embodiment condition compared to a low-embodiment condition (Lindgren and Johnson-Glenberg [2013]).

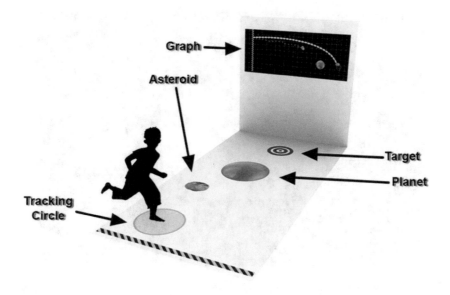

FIGURE 3.8: Schematic of the meteor installation at the Museum of Science and Industry in Tampa. (*Source*: Robb W. Lindgren.)

It is worth noting that the definition in Lindgren and Johnson-Glenberg [2013] implies that simply asking people to perform any gesture or body movement to interact with a learning system or installation is not enough to facilitate learning; rather, the work of educational designers is in identifying the proper mappings across gestures, the areas of the brain that they activate, and the desired learning outcomes. In particular, according to Segal [2011], gestures should be designed to be "congruent with the learned concept": for example, if we design an app in which a gesture is required to rotate an object on a smartphone screen, we should ask people to rotate the digital object by rotating their fingers rather than by tapping the screen (see Figure 3.9). This and other implications for design will be explored in detail in Chapter 5.

3.5   Summary: From Cognitive Science to Human-Data Interaction

In this chapter, we discussed prior work exploring the role of the body in embodied cognition and in Conceptual Metaphor Theory (CMT). We then reviewed embodied interaction and theories of learning that provide the theoretical background for understanding embodied learning. In Chapters 7 and 8), we will see how CMT can be used to frame the design of gestures and body movements for HDI installations and to explore how design choices in HDI may affect how people make sense of interactive data visualizations.

FIGURE 3.9: A Tangram puzzle in which users can rotate objects using a "congruent" gesture, i.e., by rotating their finger on the screen. (*Source*: Segal [2011].)

In the next chapter, we will review approaches and applications for designing for informal learning in museums. We will then discuss learning *about* data *through* data visualizations in Chapter 5.

CHAPTER 4

Background: Designing for Learning in Museums

While embodied interactions for learning have gained some traction in formal school settings (e.g., Enyedy et al. [2013]), non-school institutions such as museums, aquariums, zoos, and science centers (hereafter referred to collectively as "museums") are prime venues for incorporating movement-based and off-the-desktop interactions. These designed spaces for informal learning are created to engage learners with content and advance disciplinary understanding while presenting enjoyable, novel, and memorable experiences to visitors (National Research Council [2009]). Moreover, museums are increasingly attending to ways of presenting authentic data to visitors (Ma et al. [2019], J. Roberts and Lyons [2020]): in a world in which people are bombarded with data every time they surf the Internet, read newspapers, or listen to debates on TV, data are increasingly seen as "artifacts" worthy to be on display in museums. Thus, museums present an ideal setting for our exploration of embodied HDI, and we therefore focus on museums through the remainder of this book.

This chapter outlines the particularities of museum learning that must be accounted for in HDI design. First we will summarize work from museum literature on understanding museum visitors and the design context of museum spaces. We will then discuss multiple form factors for museum technology designs and introduce ways of measuring learning in museums. Before we delve into these topics, however, we must make a brief clarification on the terminology used. Through the majority of this book we have adopted the trend to use the word "informal" to describe out-of-school learning environments, following the convention that "formal" education takes place in schools and has some codified requirements regarding curriculum, duration, attendance, etc. If school equates to formal learning, then logically the learning that occurs everywhere outside school could be considered "informal" (i.e., "not school"). However, it is important to note much research on learning in museums and other out-of-school environments adopt the more precise term "free-choice" to describe the learning occurring in these spaces.

"Free-choice learning" (Falk and Dierking [2018]) refers to the lifelong, nonlinear, personally motivated experiences that constitute most of the learning we undergo throughout our lives. It is what we engage in every time we choose to read a book or article, visit a museum, browse a website, or watch a documentary on a topic of interest. It is the learning that occurs not because an instructor or authority has required it, but because we have some intrinsic motivation to do it. Free-choice learning can happen anytime and anywhere, and sometimes it can be quite formal in structure and nature. That is, the learning that we elect to do outside a classroom can be very serious and rigorous, even without the extrinsic motivation of a grade or a degree. Moreover, the attention, care, and rigor applied to the design of museum exhibits is not by any means "informal," which colloquially can imply not serious. While we don't wish to dedicate extensive space here to deconstruct the terms "informal," "free-choice," and "organic" to describe the learning interactions that occur in HDI experiences, we wanted to draw attention to their purposeful use in this chapter and throughout the volume.

4.1   Understanding People in Museums

The systematic study of museum visitors began as early as 1928 (Robinson et al. [1928]) and gained traction as a field in the 1960s and 1970s as an increasing number of evaluation studies sought to make sense of visitor behavior (Bitgood and Shettel [1996]). Museum studies researchers employ a variety of techniques in this work, including unobtrusive observations, systematic timing and tracking of visitor movements throughout a space, entry and exit surveys, and individual and group interviews (Diamond et al. [2016]).

Basic demographic features of age, education, and income give some insight to who is visiting museums, but recent decades have seen a push toward understanding museum visitors by attending to their motivations for visiting. Falk [2006] identified five identities visitors adopt in museums depending on their background, interests, and social groups: explorer, experience-seeker, facilitator, recharger, and professional/hobbyist. These motivation-driven identities are fluid and situationally dependent, and they affect the way a visitor behaves during a particular experience. For example, an adult visitor at an art museum with an out-of-town friend perhaps takes on a recharger role, seeking to use the art and the environment to relax and catch up with an old friend. That same adult will later more likely adopt a facilitator role when visiting a science center with his children. Thinking about these identities and the associated motivations and interaction needs can inform the design of HDI exhibits.

4.2   Visitor-Centered Design of Museum Experiences

In computing, we employ user-centered design strategies to create interface interactions that meet users' needs and align with their abilities and preferences. In her book *From Knowledge to Narrative: Educators and the Changing Museum*, Lisa Roberts [1997] describes a similar shift in museums from the transmission model of learning, in which a scholarly presentation of artifacts interprets content and ideas for visitors in what is essentially a lecture, to a visitor-centered model where the visitors' interests and experiences are allowed to shape the visit experience. Roberts describes the emerging role of the museum educator into a central figure in exhibition design, along with the curatorial and design teams, arguing that museums have been transformed by "the sense in which education is fundamentally a meaning-making activity that involves a constant negotiation between stories given by museums and those brought in by visitors." Rowe et al. [2002] elaborate on this theme, identifying the potential of museums to help visitors link their own individual, personal, autobiographical "little" narratives to the museum's depiction of "big" societal and cultural narratives in "mutually inter-animating ways." That is, rather than whole-cloth adoptions of a single "correct" interpretation of a historical or cultural event, visitors' own vernacular and personal narratives can contest, challenge, or elaborate the official narrative presented by the museum as part of the meaning-making experience.

The shift to the free-choice, personalized nature of the museum experience frequently leads to visitor-led learning experiences that can deviate from the museum's intentions. Alhakamy et al. [2020] describe an interactive data visualization that was designed to expose museum visitors to social data, displayed on 3D globe maps. Scaffolding questions were mostly targeted to adults and included "Does firearm ownership influence the number of murders?" and "Does female employment influence male unemployment?" Visitors, however, used this globe-based visualization in ways that went beyond the exploration of the specific datasets on display. For example, after looking at the display with her child, who was too young to engage with the framing questions of the exhibit, a parent decided to use the system to teach him geography by asking her son to name the countries on display and to point out where the United States is located on the globe. While this interaction may not have been the learning intended by the design team, empowering visitors to connect in their own way can create learning experiences more powerful and memorable than prescribed interactions.

In some cases visitors take this personalization even further, assuming the role of museum curators for their personal exhibit experience (B. Shapiro and Hall [2018]). In particular, the study in Shapiro and Hall describes how diverse groups of visitors took pictures of museum artifacts during their visit and posted them in a very structured fashion on Instagram—using hashtags and labels to classify the object in the photo—and generating engaged comments from other users.

As museum visitors—and members of the public engaging with data more broadly—are looking for their own narratives in these engagement experiences, HDI designers need to attend to the implications for design. What level of freedom and choice in data access and representation is appropriate? What scaffolding do users need to support their understanding of unfamiliar datasets? Where is the balance between curator-crafted narratives and personal meaning-making? The answers to these questions for museum and other organic data interactions are different than what is appropriate for more traditional visualizations designed for disciplinary experts. Understanding the context of social, free-choice learning environments can inform an effective visitor-centered design process.

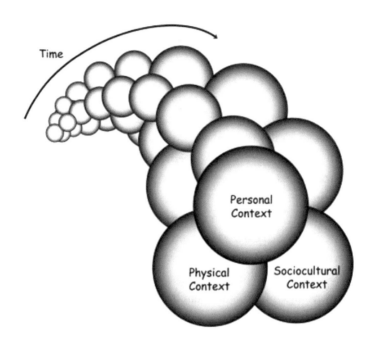

FIGURE 4.1: The Contextual Model of Learning. (*Source*: Falk and Dierking [2018].)

4.3   Museums as Social Learning Environments

Chapter 3 discussed multiple perspectives and theories on learning. These perspectives have been translated specifically to the museum context by Falk and Dierking [2018] in their Contextual Model of Learning (Figure 4.1). This model situates learning in the interactions among the personal, socio-cultural, and physical contexts across time, positing that "meaning is built up, layer upon layer." Attention to these three contexts and how they interact during a museum experience can be highly informative toward designing successful HDI exhibits.

4.3.1   The Personal Context

The personal context in the Falk and Dierking [2018] model acknowledges that what learners gain during a learning experience is inextricably tied to the personal context they brought into the experience—prior knowledge, experiences, motivations, identities, etc.—as well as where they go next. The authors illustrate this concept through an anecdote of two colleagues visiting the Smithsonian Natural History Museum while attending a conference in Washington, D.C. Researchers observed the pair during their 90-minute trip to the museum and interviewed them as they exited about what they saw and remembered, and then researchers interviewed both subjects again five months later. The follow-up interview showed striking differences between the subjects' memories of what was interesting and important about the visit, noting that these differences tie back to their previously stated individual personal and professional interests. As Falk and Dierking note:

> Perhaps most startling is the fact that these demographically nearly identical individuals (well-educated, professional—both children's textbook editors—white women in their late twenties living in the same Chicago community) visited the museum on the same day, saw the same exhibitions for exactly the same amount of time, even viewed and discussed among themselves some of the same specific elements, and yet what they learned/remembered was totally different.

Their example is hardly an outlier, and it is certainly not unique to museum learning. Recall from Chapter 3 that the basic tenet of constructivism is that we are constructing new knowledge based on our existing knowledge. The more existing knowledge and interest we have on a subject, the more connections we can make to new information we encounter and therefore the more likely we will retrieve that information later. Research in data interpretation has shown that people naturally draw on personal understandings and experiences to make sense of representations (Peck et al. [2019]), and the "frames and slants" in the wording of titles of visualizations impact interpretation and recall (Kong et al. [2018]). This inherent tendency for learners to construct their meaning based on preconceptions means that the exhibit designer's challenge is to find a way to maximize the potential for visitors to form these connections while minimizing instances of misinterpretation. As will be discussed, the personalization of an experience afforded by embodiment can help achieve these goals, particularly in collaborative, social data exploration (J. Roberts and Lyons [2020], J. Roberts et al. [2014]).

4.3.2   The Sociocultural Context

The sociocultural context in Falk and Dierking's [2018] model posits that learning is mediated by social interaction, as "learning, particularly learning in museums, is a fundamentally social experience." This perspective follows the Vygotskian view that humans virtually always learn through dialogues, asking questions, and negotiating meaning. In practical terms, the sociocultural perspective can be operationalized as: a social group or ensemble is engaged in an activity; this activity is collaborative and is informed by the individuals who comprise it and yet the activity reciprocally informs the individuals/group; the social activity is mediated by tools, signs, people, symbols, language, and actions (Ash [2003]).

Falk and Dierking [2018] point out that museums are ideal environments for capitalizing on the sociocultural context for learning because unlike formal environments like classrooms, "in the real world, ... if you do not know the answer to something you want to know about, you ask for help, read about it, or in some way seek out ways to maximize your zone of proximal development. Free-choice learning in general and museum learning in particular are commonly marked by some sort of socially facilitated learning." Atkins et al. [2009] agree, saying, "we view one of the richest forms of learning in a museum to be evident in the patterns of discourse and activities that groups engage in—such as labeling, theorizing, predicting, recognizing patterns, testing ideas, and explaining observations. These patterned activities provide a structure through which visitors construct scientific ideas and, in doing so, learn what it means to participate in scientific activities."

Whether accompanied by friends, family members, or an organized group such as a school field trip, visitors share their experiences with each other through dialogue and co-construct knowledge based on individual and shared prior knowledge and experiences. Paris [1997], as cited by Packer and Ballantyne [2005], suggests five ways in which social interaction facilitates museum visitor learning:

1. People stimulate each other's imaginations and negotiate meaning from different perspectives;
2. The shared goal of learning together enhances motivation;
3. There are social supports for the learning process;
4. People learn through observation and modeling; and
5. Companions provide benchmarks for monitoring accomplishment.

Embodied designs make human-data interactions visible, granting co-visitors awareness of each other's actions as they move in a shared space. This awareness promotes social interactions and opens opportunities for socially mediated learning as visitors talk to each other about data as they explore it (J. Roberts and Lyons [2017b]).

4.3.3   The Physical Context

The third context embraced in the Contextual Model of Learning is the physical context, what Wertsch [1998] (citing Kenneth Burke [1989]) calls "the scene." This is the "container" where the mediated action takes place, and can be scoped to "a great variety of circumferences" (Wertsch [1998]). In the context of museum exhibits, designers, researchers, and educators have explicit control over the physical context. The aesthetic feel of a museum will influence how most visitors behave there; it is thus not surprising that many embodied HDI exhibits can be found in high-energy interactive science centers where the physical context invites movement and play.

As a final note on the Contextual Model of Learning, though the personal, sociocultural, and physical contexts are presented separately, the three contexts are "not really separate, or even separable," nor are they "ever stable or constant" (Falk and Dierking [2018]). They are part of a moving and interconnected system in which current conditions are mediated by past experiences. Looking at only a snapshot of an individual episode in a person's life to understand their learning is "woefully inadequate" (Falk and Dierking [2018]. Thus the model incorporates time through the stacking of layers representing past states (an effect that in some circles has given the model in Figure 4.1 the loving moniker "the Worm").

This complexity of understanding how to design data interactions in these spaces is further complicated by the reality that visitors often attend in heterogeneous, inter-generational groups. As will be discussed next, designing for multiple ages and abilities simultaneously presents its own challenges.

4.4   Meeting the Needs of Inter-Generational Groups

We established above that even demographically similar museum visitors can have entirely unique visit experiences based on their own personal background and interests, and yet it is crucial to point out that many visitors come in quite heterogeneous groups, in particular families with adults and multiple children of different ages. This scenario is very different from what happens in formal learning, where students are typically grouped by age in classes and have some curriculum history in common. The diversity of museum groups offers learning opportunities but also requires a specific attention to designing for this mixed crowd of visitors.

In the book *What Makes Learning Fun?* Perry [2012] describes the process of re-designing Colored Shadows, an installation at the Children's Museum in Indianapolis that allowed visitors to combine lights of different wavelengths in order to produce new colors. Perry's [1993] work is centered on the idea that a "successful" trip to a museum means that visitors: (1) have an enjoyable, social experience and (2) are able to learn something.

Although these two principles may seem trivial at a first glance, designing exhibits and installations that fully support them is not a trivial task. For example, Perry [2012] describes an interesting anecdote of what may happen because people of different age groups visit a museum at the same time:

One carefully dressed father approached the exhibit with his young son, who was probably in the fifth grade. They weren't talking very much as they wandered around the gallery; nor did they seem particularly excited or animated. But as they drew closer to Colored Shadows, the father brightened and became visibly interested. As with most inter-generational groups visiting this exhibit, the younger visitor approached first and began waving his hands over the tabletop, making brightly colored shadows. The father excitedly began reading the Parent Information interpretive text that explained—in quite a bit of detail—color and light theory. Although initially intrigued by the colored shadows he was creating, the boy quickly became bored and started to leave the exhibit. (p. 20)

This anecdote describes the tension in designing exhibits for different age groups: determining for whom an exhibit should be designed. Sometimes exhibits target learning goals for specific age groups, but this strategy can leave others disengaged (e.g., exhibits at children's museums that are very fun for children but quite tedious for parents). For example, we found particular challenges in calibrating embodied interaction controls for CoCensus, a whole-body interaction census data museum exhibit (Cafaro et al. [2013], J. Roberts et al. [2014]). Adult visitors were disinclined to carry out larger, more physical control actions like jumping and often looked self-conscious doing even smaller arm-waving gestures. Children, on the other hand, hyper-emphasized movements, jumping and waving well beyond what was required for the exhibit controls (several times, in fact, wearing through the floor covering that marked the temporary interaction area during pilot testing; we ultimately replaced it with a more robust finalized design to withstand the wear and tear).

Still, designers have uncovered strategies to engage diverse age groups. Perry [2012] recommends creating interpretative labels with alternative prompts and questions to target different age groups, for example, sections with Parent Information that describe what parents could discuss or do with preschoolers when exploring an exhibit or installation, along with simpler text that K-12 children can read by themselves during a school field trip. Lyons et al. [2010] review multiple ways technologies can be designed to frame activities and provide just-in-time resources for parents looking to support their children's learning while avoiding the common trap of taking over the complex cognitive tasks themselves and involving children only in physical and logistical tasks (Schauble et al. [2002]).

Inquiry Games for family groups (Allen and Gutwill [2009]) and student groups (Gutwill and Allen [2012]) provides an example of heterogeneous groups engaging in activities together through structured games. In this study, researchers gave visitors a crash course of inquiry methods on the museum floor by providing guides for conducting inquiry-based activities through inquiry skills. In pilot work families were taught a sequence of six skills (e.g., exploration, question generation, metacognitive self-assessment; see Allen and Gutwill [2009] for full description) at an exhibit before they were asked to apply those skills at a novel exhibit. Observations demonstrated a propensity to forget the prescribed sequence and abandon some skills, prompting researchers to refine the inquiry skills to *proposing actions* and *interpreting results*. It was determined that "these two skills complement students' natural exploration activity at exhibits, are intellectually accessible to diverse groups of students . . . and are simple enough for students to understand quickly and remember easily." These skills became the foundation of two games: Juicy Questions and Hands Off (Gutwill and Allen [2012]). While these games, particularly Juicy Questions, were found to positively impact visitors' engagement and learning, the required investment of time and resources for the researchers to teach the games to visitors makes widespread use of these tactics impractical.

While few exhibits are likely to engage every visitor and all visitor group combinations, some design strategies—and in particular, some embodied and off-the-desktop technologies—have shown promise in supporting mixed-age and mixed-interest groups. A review of design strategies that begin to address these challenges is presented next.

4.5 Choosing Technology Designs for Visitor Interactions

Why should we choose embodied interaction over other, more common technologies like touch-screens and mobile devices? We know from decades of research on the design of learning environments and from studies specifically situated in museums that seemingly small design decisions can have large impacts on the dialogue and interactions taking place around an exhibit. A variety of technology-based strategies have been used for engaging visitors with content, including screen-based interactives, handheld games for groups, multi-user touchscreen devices such as tabletops, and spatially indexed interactive experiences. Here we will discuss two popular form factors— screen-based and handheld interactives—to provide an overview of how these technologies have been integrated into museum experiences and their affordances in supporting learning.

4.5.1 Screen-Based Interactives

Many exhibits incorporate computer-based displays to provide extra information, beyond what can be presented on a traditional printed label. These screen-based interactives offer supplemental text and video, quiz visitors on domain knowledge, and present simulations illustrating key concepts. One early such example is the Sickle-Cell Counselor (Bell et al. [1994]), a computer program placing visitors in the position of a genetic counselor advising couples on the risk of genetic disease in their offspring. Visitors can perform simulated lab tests and ask scripted questions of experts by viewing pre-recorded videos in order to make their recommendations. This goal-based scenario design provides users with an easily recognizable goal (to advise the couple) and opportunity for deep engagement with the data through authentic practices (lab tests and consultations with experts).

Heath and Vom Lehn [2008] note that the single-user input design exemplified by Sickle Cell Counselor is prevalent in many computer-based museum interactives, which subscribe to a model that "prioritizes the single user and disregards the socially organized interaction that underpins the use of technologies... It is as if the design of exhibits presupposes a neutral domain that consists of a series of isolated individuals and individual actions, who at best are prepared to wait their turn and if necessary become a passive audience."

Hornecker [2010] notes a similar sense of isolation among visitors using Juroscopes, media-augmented telescopes at a natural history museum. When an individual looks through the Juroscope at a dinosaur skeleton, they see 30-second animations in which the skeletons grow organs and skin to look like live dinosaurs in their natural environments—moving, feeding, and hunting, including charging at the visitors themselves. These popular individual devices provide an immersive experience, causing some visitors to cringe and pull back from the viewfinder when the dinosaur "notices" them. But the isolated experience makes conversations among visitors disconnected. As Hornecker notes, "Being virtually alone adds to the feeling of a direct and personal experience. It also means it is difficult to share." Dialogue around these devices was compared with a barrier-free version of the same animations that utilizes a large, angled screen. These screens were intended for wheelchair users and children too short to reach the Juroscopes but came to attract larger crowds of visitors of typically 6 to 15 people who were "not just waiting, but actively watching and commenting, scaffolding and negotiating use of the lever mechanism (for selecting the dinosaur to be animated)" (Hornecker [2010]). The barrier-free versions lost some of the excitement of full immersion but gained social interactivity.

Similar work at the Field Museum of Natural History in Chicago, IL addressed how multi-user touchscreens could engage users with artifacts behind glass. Each object case in an exhibit on the history and culture of China was supplemented with a touchscreen providing information about each artifact on display, including supplemental text, images, and videos. The screens were single-touch interfaces but were wide enough for multiple users to access the content together. A study testing multiple interface designs of these screens (see Figure 4.2) found significant differences in how people used the screens and discussed the related content depending on whether they used a version featuring a large curiosity-inducing question, a graphic-based interaction, or a formal table-of-contents-like view (J. Roberts et al. [2018]). Each screen design had different affordances for supporting dialogue among companions.

Other research supports the idea that simply enlarging the display for some technology-based exhibits removes one of the isolating components described above: a large display provides a shared output, letting multiple visitors engage more easily (DiPaola and Akai [2006]). For example, Bao and Gergle [2009] manipulated visual display sizes (large wall-sized vs. small desktop) and tasks (object identification and narrative description) and found significant differences in user references. Specifically, they found that presenting visual information on large screens was associated with significantly increased usage of deictic pronouns (e.g., "this," "that") when users produced open-ended narrative descriptions of the visual data. From these findings, they posited that producing narratives about information presented in large displays may "yield a more immersive experience, or greater sense of presence, that translates into measurable differences in language use" (Bao and Gergle [2009]). An additional strategy is to distribute the control of the screen across multiple visitors, but that requires careful planning on which functionality can be distributed and on the timing of the coordinated actions that allow for distributed control (Clarke et al. [2021]).

The manner in which the shared activity is framed also has significant impact on groups' interactions. For example, Atkins et al. [2009] recorded visitor interactions around two versions of

FIGURE 4.2: The original table-of-contents design (left) presented indexed artifacts in the formal style typical of traditional object-based museums. The graphic-based interaction (right) pulled a slideshow of characters in different styles of writing into an interactive timeline that allowed users to explore when each was used and which styles were in use concurrently.

a heat camera exhibit at a science center (see Figure 4.3). The functionality of the exhibit was the same in both conditions—a screen showed how objects look through an infrared camera—but one version provided multiple styles of mittens for visitors to try on and included brief explanatory text inviting them to observe differences in how each trapped heat. The alternate version had no such framing activity and just allowed for open explanation. Though in each case the exhibit did the same thing, what the visitors tried to do with it varied greatly. The "mittens" version of the exhibit yielded more classroom-like talk, in which one visitor, typically a parent, tried to instruct others about the lesson and what people were "supposed" to do and learn with the exhibit; by contrast, the "no mittens" version prompted exploration and "creating data" by spontaneous investigations of other objects in the space (Atkins et al. [2009]). The little bit of framing led to qualitatively very different interactions. Neither type of dialogue is inherently good or bad, but the differences need to be kept in mind depending on the kinds of thinking and talk an exhibit is intended to promote.

While screen-based displays can provide unprecedented amounts and forms of interactivity and information, designers must attend to the inherent tradeoffs between highly immersive but exclusive individual experiences and group engagement in a less personalized interaction. These tradeoffs become even more pronounced when scaling down to handheld mobile interfaces.

FIGURE 4.3: Users experiment with a heat camera exhibit similar to that tested in Atkins et al. [2009].

4.5.2 Mobile Applications

Museums have increasingly integrated smartphones and mobile technologies to their offerings for visitors (Schultz [2013]). QR codes are now frequently found near museum artifacts: they supplement the exhibit label by delivering additional interpretation to the visitors' mobile phones (Pérez-Sanagustín et al. [2016]), or even direct robot tour guides through the exhibit space (S. J. Lee et al. [2014]). Klopfer et al. [2005] describe a mobile-based, collaborative game that was tested at the Boston Museum of Science. Visitors were challenged to work in groups of six to identify a thief that had stolen a museum artifact by interviewing virtual characters across the museum rooms. Augmented reality applications have also been designed to create novel exhibit experiences. For instance, Thian [2012] uses a mobile application to provide augmented reality content when yellow markers are detected through the exhibit space (see Figure 4.4).

The use of smartphones in social spaces like museums is, however, controversial: mobile applications can capture peoples' attention towards a small screen, limiting the social experience that they would otherwise enjoy while visiting a museum (and, as we discussed early in this chapter, this social aspect can be crucial for informal learning). Klopfer et al. [2005] note that museums have employed audio tours and handled devices to provide visitors with information, but there is a lack of focus on how to use these technologies to promote collaboration. In 2003, Hsi [2003] conducted a study at the Exploratorium, a science museum in San Francisco, CA, with 15 museum

FIGURE 4.4: Visitors use a mobile application to recognize markers in an exhibit on Terracotta Warriors at the Asian Civilisations Museum in Singapore. (*Source*: Thian [2012].)

visitors (college students, museum members, and school teachers). Participants were given a hand-held device that received exhibit-related web pages when the user was in the proximity of RFID beacons. In follow-up interviews, participants described feeling "a sense of isolation" from the experience, complained that they were spending more time interacting with the handheld device than with the exhibit, and reported that they did not even notice other people around them. The work in Cafaro et al. [2013] takes a similar perspective to explain the benefits of embodied interaction in museums: "Modern technologies, including handhelds, may tend to isolate us from one another." On the other hand, embodied interaction techniques (such as tangible and full-body) are uniquely suited to support the social environment of museums, because they fully use the (social) space in which the exhibit is located.

This potential does not mean, however, that full-body interactive systems will be universally better. In one counter-example, Petrelli and O'Brien [2018] compare tangible interaction vs. smart-phone applications as a way to control museum installations and noticed that mobile apps limit the engagement with installations, but they do favor mobility through the museum space. Specifically, researchers observed 76 visitors interacting with an exhibit about a defensive wall built by the Nazis during World War II. Using a classification introduced in Véron and Levasseur [1989], visitors were labeled as Ants if they tended to get close to exhibit artifacts and observe them and as Fishes if they spent most of their time close to the center of the exhibit. The researchers observed that Fishes mostly used a phone application to interact with the exhibit, while Ants generally preferred to use tangible objects to control the installation. Thus, the choice of the best interaction modality may depend on the personality of each user (Ant vs. Fish).

J. Roberts and Lyons [2017b] likewise find that full-body controls were not necessarily supe-rior in supporting visitors engaging in learning talk (a useful marker of learning in museums, as described next) during data exploration. In a study comparing full-body interaction versions of the CoCensus exhibit with an identical display controlled through a handheld tablet interface, it was found that even when controlling for duration of the interactions and number of control movements, users in the handheld conditions produced significantly more learning talk. While the findings (dis-cussed more in Chapter 8) were counter-intuitive to the notion that embodiment would support social learning, they present a much-needed word of caution when considering novel interaction design for complex information displays. Excitement for the novelty should not be allowed to trump attention to the intended learning outcomes. Measuring this learning, however, is still an open challenge in free-choice museum environments.

4.6   Measuring Learning in Museums

So far, this chapter has provided an overview of theories and examples of designs for learning in informal learning environments. The crucial question before closing this chapter is the question that hangs over every educational design: How do we know whether and what people have learned?

Creating valid and reliable measurements of learning is a challenge in any learning environ-ment, and the free-choice, open-ended, and self-directed nature of museum visits make assessment particularly difficult. The knowledge-based exams often relied upon by formal schooling can rarely tease apart what was learned during a museum visit from what prior knowledge the visitor already possessed. In their *Practical Evaluation Guide: Tools for Museums and Other Informal Educational Set-tings*, Diamond et al. [2016] lay out a series of strategies and considerations for measuring museum learning. For example, familiar measures like surveys and interviews can elicit knowledge retention and even implicit memory, while asking visitors to create personal meaning maps on the topic or sketch the exhibit they just visited can provide more open-ended opportunities for visitors to share the aspects of a visit they have most front-of-mind.

Visitor observations—either cued observations, where visitors are asked if they can be followed, or uncued, surreptitious, surveillance—are commonly employed to reveal behaviors that can be used as a proxy for learning. During a museum visit, people engage in "performance indicators" and "significant behaviors" (Borun et al. [1996]) or "engagement behaviors" (Packer and Ballantyne [2005]) such as looking at displays, reading text, asking and answering questions, and engaging actively with exhibits. Such behaviors give a picture of what visitors are doing during their trip, and some of these behaviors have been linked to learning. For instance, Borun et al. triangulate video and audio recordings and post-visit interviews to link observable behaviors of family groups on a museum visit with the groups' learning about the exhibit content. They found the frequency of certain "significant behaviors" to be a distinguishing factor between family groups in successive learning levels. These significant behaviors were:

- Ask a question;
- Answer a question;
- Comment on the exhibit, including explaining how to use the exhibit (for interactives);
- Read the label aloud;
- Read the label silently.

These categories, along with movement tracking and counting heads, are common in the museum learning literature, with some adaptations. Diamond et al. [2016] caution, for example, that "you may not be able to verify whether a visitor read the exhibit label unless he or she read it aloud" and therefore advise the use of "look at label" instead of "read the label silently." Regardless, observations can give some insight to how the groups are interacting with each other and the exhibit and may be indicators of the learning occurring in the groups.

4.6.1   Learning Talk

If learning is socially mediated, then dialogue among visitors can provide insight to how visitors are learning. Allen [2003] dubs productive, on-task dialogue between visitors as "learning talk." Often analyzed at the group rather than individual level, learning talk analysis typically draws upon categories for affective, cognitive, and psychomotor learning. Allen's coding scheme divided 16 subcategories of codes into five categories: Perceptual, Conceptual, Connecting, Strategic, and Affective. These codes and subcategories were emergent based on the dialogue in order to capture the nature of the conversations and were iteratively refined to adequately capture the evidence of visitors' learning. Atkins et al. [2009] use a scheme bearing resemblance to Allen's codes, but include the categories Navigation, Creating and Noticing Data, Experimenting, and Affect, each with two to eight subcategories.

Typical learning talk analysis contributes to primarily qualitative findings, but quantification of learning talk can provide valuable insights (Chi [1997]). When applying a coding scheme derived from Allen [2003] and informed by graph interpretation literature by Friel et al. [2001] to analysis of visitors' data interpretation using the CoCensus exhibit, we used a combination of *simultaneous* and *magnitude* coding to compute learning talk scores per user session (J. Roberts and Lyons [2017a,b]). We coded each segment of learning talk uttered by pairs of visitors to CoCensus with multiple codes and weighted those codes according to the learning objectives of the exhibit. The numerical weights were summed into a session score for each visitor group with which we could quantitatively compare multiple versions of the exhibit to pick a winning design. More on CoCensus, and how the analysis of learning talk contributed to our understanding of the implications of embodied exhibit design, will be discussed in Chapter 8.

4.7   Summary: Doing HDI in Informal Learning Settings

Informal learning settings, and in particular museums, are the application scenario for the HDI installations that we discuss in this book. In this chapter, we provided an overview on relevant museum literature. We then discussed theoretical and practical considerations for designing museum learning experiences and introduced metrics for assessing their impact on learning. In the next chapter, we complete our review of the background literature by summarizing work that explores the use of data visualizations to facilitate meaning-making, particularly in informal learning settings.

CHAPTER 5

Background: Visualizations to Support Learning

The previous chapters examined the theoretical underpinnings of embodiment, outlined relevant learning theories, and described key contextual factors particular to the museum learning environment. Before we move on to the design of embodied HDI experiences in museums, we turn our attention here to the use of visualizations as a learning tool. This chapter will explore the nexus of visualization and learning research with a focus on how visualizations have been adapted for non-experts. In particular, we focus on graphs and map-based visualizations, which are particularly used in formal and informal learning settings.

5.1   Visualization in Public Spaces

Shifts in data collection methods, storage procedures, and expectations for transparency in recent decades have resulted in unprecedented amounts of data being available to the public for download and exploration. However, as a large body of research on data literacy has shown, making data available is not the same as making data accessible. While visualizations have potential to aid interpretation regardless of the content being depicted, it is widely acknowledged that the purpose and audience must be kept in mind when designing a graph or visualization (Few [2009], Fischer et al. [2005], Glazer [2011], McCabe [2009], Shah and Hoeffner [2002]). Graphs designed for novices and children, for example, must take into account cognitive load and their available working memory resources by reducing the difficulty in keeping track of graphic references (Friel et al. [2001], Plass et al. [2009], Shah and Hoeffner [2002]). Furthermore, Libarkin and Brick [2002] point out that "cognitive skills must be acquired to assist in interpreting visual representations of actual phenomena. These skills are not necessarily a natural consequence of exposure to visual communication, and scaffolding between verbal and visual modalities may be an integral component of effective communication." Essentially, we cannot assume that simply providing a graphic to learners will ensure that they can and will interpret it, much less interpret it accurately and critically.

Neither can we assume that learners will bring a common set of schemata or interpretive tools to the display. As Shah and Hoeffner [2002] point out, "People's knowledge of the content in graphs has an influence on their interpretations of, and memory for, data … This is especially true for novice graph viewers who often do not have the graph schemata necessary to overcome the strong influence of their own content knowledge." Friel et al. [2001] agree, noting, "The graph reader's situational knowledge may interrupt her work on the cognitive, information-processing tasks performed in interpreting the graph." Even in a classroom where students share a curriculum, we cannot assume shared schemata and prior knowledge when approaching visualizations. This is especially true in a museum, where a wide variety of visitors with varied education and experience mingle. Visualization research has shown that even data fluent users are prone to bias (Gotz et al. [2016], Wall et al. [2019]), and domain expertise is not necessarily sufficient for exploratory analysis on complex datasets (Perer and Shneiderman [2008]). Users face challenges both in translating existing data in tables, maps, and graphs into words (i.e., describing what a representation means) and in extrapolating to the larger themes and implications of those data, alone and as part of a complex system (Fischer et al. [2005], Friel et al. [2001], Glazer [2011], Libarkin and Brick [2002], Shah and Hoeffner [2002], Uttal [2000]).

Much work has been done to understand how people understand and interpret scientific graphs and maps in classroom settings, where teachers guide students through sequenced activities to bridge students' prior knowledge and build motivation to sustain prolonged inquiry (e.g., Edelson et al. [1999]). Data science curricula often aim to create a transparent relationship between the learner and data at all phases of the data life cycle (Berman et al. [2018]) or by engaging students in designing, collecting, and analyzing their own datasets (R. Lee and Drake [2013]).

In informal learning environments (i.e., outside of work or school), however, learners do not have the privilege of these prolonged interactions, and moreover, users are likely neither domain experts nor visualization experts. What, then, are the challenges they face in human-data interactions?

Peck et al. [2019] investigate this question via a study of residents of rural Pennsylvania. Semi-structured interviews explored how 42 residents made sense of unfamiliar data representations. Their analysis used a diverse representation of styles and sources of data representations and focused on where participants focused their attention while completing a task of sorting a set of ten representations from most useful to least useful. Follow-up questions probed the participants on their choices, particularly when two similar representations (e.g., two line graphs) were ranked differently. Sources of each representation were only revealed after the initial ranking, and after learning the sources participants were invited to re-rank the graphs if they chose to. The grounded theory analysis (Charmaz and Belgrave [2007], Strauss and Corbin [1997]), in which codes are derived from the data rather than pre-generated based on prior literature, revealed the deeply personal nature of data interpretation, noting, "The most recurring theme in our analysis were decisions framed or driven by personal experience." This personal nature of non-expert data interpretation practices, while potentially challenging, suggests some serendipity with the design of informal education spaces. Museums and other free-choice learning environments already focus on fostering connections from visitors' prior knowledge and ideas to the novel content presented in an exhibit. Data interpretation may require some new techniques and scaffolding, but capitalizing on the personal nature of data interpretation is a promising direction for museum design.

It is not surprising, then, that work in museums has begun exploring data visualizations for conveying concepts to visitors in topics ranging from phylogeny (Davis et al. [2013]) to marine microbes (Ma et al. [2012]) to census data (J. Roberts et al. [2014]). Early work recognizing the specific nature of information visualization for a museum audience observed visitors using a cut section visualization of discourse on artist Emily Carr's life and work over time in an exhibit called EMDialog (Hinrichs et al. [2008]). Through an observational study the research team noted the need for museum-based visual displays to accommodate a variety of interaction lengths, exploration styles, and visitor groupings.

Ma et al. [2020] conducted a deeper exploration of visitor groups through observations and think-aloud protocols with Exploratorium visitors using the interactive Plankton Populations tabletop exhibit. Researchers coded think-aloud dialogue from 56 dyads of visitors to identify decoding comments and data interpretations. The team found that decoding was "an ongoing act of construction" throughout visitor interactions, and the median time it took for dyads to make their first interpretation comment was 43 seconds, with the median before the first *correct* interpretation comment being 53 seconds. They contextualize these numbers, stating, "Considering the museum context, where the total holding time is measured in seconds, 43 seconds is a long time for a visitor to arrive at his/her first data interpretation. As a point of reference, the holding times for exhibits in an earlier Exploratorium life sciences collection ranged from 12 to 149 seconds" (Ma et al. [2020]).

These findings echo troubles identified by Börner et al. [2016] in a multi-phase study of visualization literacy among science museum visitors. The research team conducted interviews with 273 visitors at three science museums in the United States. Visitors were shown a series of five visualizations and asked about their familiarity with each data presentation, how they thought it should be read, what it should be called, and what types of data or information made sense for each type of visual. While the study team found that visitors overall were interested in the visualizations, they faced significant challenges in interpreting unfamiliar representations (Börner et al. [2016]).

Still, data presentations are increasingly incorporated into museum exhibits, and exhibit and visualization designers must endeavor to improve their accessibility and readability for museum audiences. The remainder of this volume will argue the potential for embodiment to support that aim, but first we look to some of the decoding processes that learners undertake when approaching visualizations.

5.2   Graph Interpretation

Data reasoning goes beyond reading the details of the representation (e.g., naming locations and describing key features in a map) to deeper engagement with the data, such as posing questions and hypotheses, inferring relationships among variables, connecting data to external knowledge, and making predictions. Much work has been done to understand how people understand and interpret scientific graphs. Shah and Hoeffner [2002] provide a review of empirical studies on the matter, describing three major component processes of graph interpretation: (1) encoding visual information and identifying important visual features; (2) relating visual features to the conceptual relations they represent; and (3) determining the referent of the concepts being quantified and associating those referents to the encoded functions. The three main factors affecting these interpretive processes, the authors continue, are (1) the characteristics of the visual display; (2) the viewer's knowledge of graphical schemata and conventions; and (3) the content of the graph and the viewer's prior knowledge and expectations about the content.

Friel et al. [2001] identify three kinds of behaviors involved in graph comprehension that can be extrapolated to serve as a taxonomy of data reasoning statements about a data map or any other visualization.

At the lowest level of the taxonomy are *translation* statements. These utterances merely change the form of communication, describing what is being displayed without any additional reflection or interpretation. Translation statements may be useful in grounding oneself and one's companions in the environment to get a handle on the display, but they do not indicate engagement or deeper meaning-making beyond basic comprehension.

The next level of the taxonomy consists of *interpretation* statements. These statements look for relationships among two or more pieces of information and involve rearranging and sorting important from less important information as well as identifying patterns. Examples of interpretation statements could include those comparing two pieces of data, identifying changes in a dataset over time, or contrasting two geographic locations.

The highest level in this framework is labeled *extrapolation and interpolation*. Statements in this category extend the reasoning illustrated in interpretation statements by integrating outside knowledge or extending the reasoning beyond the representation. Examples include predictions, hypotheses, and inferences about the data.

5.3   Maps as Reasoning Tools

Geographic Information Systems (GIS) maps overlay spatially referenced data onto a geographic area and are among the most common forms of informal data presentations. These data maps visualize relationships and contrasts in striking ways in order to allow users to reason about the presented data. When used properly, GIS can assist with the recall of information (Rittschof and Kulhavy

[1998]) and facilitate spatial reasoning and higher levels of thinking (Lloyd [2001], Marsh et al. [2007]). Though data maps are one of the oldest and most popular forms of data visualizations and proliferate online, in textbooks, in the news, and in museums, people still have a hard time interpreting them correctly, and with a critical eye (Monmonier [2018]). Maps can be particularly powerful tools for decoding data because "maps affect how we think about spatial information; maps may lead people to think about space in more abstract and relational ways than they would otherwise. For these reasons, maps can be construed as tools for thought in the domain of spatial cognition" (Uttal [2000]). However, novice map readers need to be taught how to effectively utilize these tools, and especially to understand how maps "lie" and can easily distort data in order to convey a particular message (Krygier and Wood [2016], Monmonier [2018]).

Much literature related to the interpretation of maps focuses on wayfinding (e.g., Beheshti et al. [2012]), but some researchers are concerning themselves with the map as a spatial cognition tool. Uttal [2000] reviews the utilities of maps for helping us conceive of the world beyond our immediate experiences, for making different kinds of information perceptually available, and for highlighting abstract spaces. Ress et al. [2018] describe design strategies and functionalities for interactive maps that help visitors of historic places in navigating complex historic narratives that span across time and space.

Some work has addressed the incorporation of geospatial data into classroom inquiry activities. Groups of students in both middle school and undergraduate contexts have used a web-based GIS for inquiry into migrations and neighborhood change (Radinsky et al. [2012]). While students were able to use the GIS to support their narratives with some success, these interventions consisted of multi-week units in which students' inquiry skills and content knowledge were built through prolonged engagement. Discussion and reflection were guided by teachers who were able to appropriately scaffold the students in supporting claims with evidence and considering multiple sources of information. Still, the undergraduate students who were given access to the full corpus of data and a wide range of manipulations of the maps were found in some cases to misrepresent the data in their presentations, for example conflating census variables. Middle school students were given a pared-down version of the interface presenting only a small set of census variables with limited data manipulations available in order to scaffold their explorations. In short, novice users (i.e., students) required a great degree of support in order to engage with the data.

Similarly, Edelson et al. [1999] discuss efforts to implement technology supported inquiry learning in classrooms through a multi-week inquiry unit around global warming utilizing custom-built scientific visualization software. In their work designing an interface to engage students with weather data, Edelson and colleagues discovered that merely giving students data maps did not immediately engage them with the content. The team had assumed that having access to authentic data would be motivating. Instead, they found that many students lost interest after looking up data on the date and location where they were born. Iterations of the curriculum incorporated an introductory activity in which students created their own color-coded map representations of their predictions about weather patterns. This activity served as a bridge between something familiar and appealing (coloring) to the task at hand (interpreting weather visualizations) and helped the students later interpret similarly visualized representations of actual data. Such bridging activities "can address motivational, accessibility, and background knowledge issues" (Edelson et al. [1999]). Through multiple iterations of the software and curriculum design, Edelson et al. discovered the importance of fostering student motivation, carefully sequencing activities to bridge students' prior knowledge, and developing supports for documenting new information in order to sustain prolonged inquiry.

These studies suggest the power of GIS and interactive data representations more generally as tools for supporting learning but highlight the challenges of engaging learners even in a formal classroom environment with a teacher and extended curriculum present to facilitate and guide the learning. In an informal, unstructured museum exhibit experience, these challenges are amplified. Yet the social and playful nature of museums may provide affordances for collaborative engagement as well. Geovisualizations, if designed in ways that afford collaborative investigations, have great potential to support dialogue (MacEachren [2005]). MacEachren and Brewer [2004] coin the term "geocollaboration" as "visually enabled collaboration with geospatial information through geospatial technologies." The potential for embodied interaction to meet these challenges will be explored in the remainder of this volume.

5.4   Summary: Data Representations as Tools for Learning

All visualizations require decisions to be made about what will (and won't) be visualized and how it will be presented. These decisions fundamentally impact how the visualization will be interpreted (e.g., McCabe [2009], Monmonier [2018], Plass et al. [2009], Shah and Hoeffner [2002], Uttal [2000]). Providing users with agency in selecting and manipulating the representation affords them the opportunity to see and play with data in new ways; rather than experiencing a limited representation demonstrating a narrow, curated narrative, users can explore different representations, each with unique affordances for data interpretation and reasoning. This capacity for open exploration matches well to the nature of free-choice learning environments, where learners are supported in choosing their own trajectory and exploring according to their own interests (Falk and Dierking [2018]). Entirely unconstrained interactions, however, can be detrimental for novice users, who can be overwhelmed and get lost in what Marsh et al. [2007] call "buttonology," in which their focus is on clicking buttons to get things to happen rather than on content reasoning. Particularly in an informal learning environment like a museum, constraints on a visualization's interactivity can be key for appropriately scaffolding the learning experience.

CHAPTER 6

Designing Engaging Human-Data Interactions

In this chapter, we narrow down the focus on human-data interaction in museums. The overarching theme is that museum visitors typically spend less than two minutes with an exhibit, even one that they really like (Sandifer [1997]; Ma et. al [2020]). The brevity of these interactions, in combination with the distinctly different—or even absent—motivations for engaging deeply with the data, calls for unique design requirements. We discuss how embodiment and novel, engaging interactions can provide extrinsic motivations to (literally) jump in, and we describe strategies to lure visitors into the interaction with HDI systems.

6.1   Engaging Museum Visitors in Data Exploration

Ma et al. [2020] explain that a first critical step in engaging visitors with data is to facilitate *decoding*, i.e., "the process by which visitors map the visual elements within a visualization to the data and data relationships that they are meant to represent." While observing people interacting with a data visualization representing marine plankton at the Exploratorium in San Francisco, they noticed that it takes visitors at least 43 seconds to come up with an interpretation of the data on display. They also found that some commonly used visualization strategies, such as including secondary data or using color coding for representing different data types, may further delay interpretation.

The challenges for engaging people in HDI, however, are not limited to the design of the data visualization. Even the delivery medium (in most cases, the display) can be an obstacle for the interaction. Figure 6.1 shows an emblematic example from a formative study that we conducted at another science museum (Discovery Place Science, Charlotte, NC) in 2017. A museum visitor (who previously tried unsuccessfully to interact with a full-body installation) decided that the system was not working, left, and, after coming back with a book, took a seat on the projector stool in front of the screen, inhibiting other people from accessing the interactive display and the nearby installations.

6.2   Challenges for Interactive Public Displays

How can this happen? After all, we had spent months designing and implementing a motion tracking system and an interactive data visualization, and we had already conducted a few in-lab studies

FIGURE 6.1: A museum visitor sits on the projector stool in front of the screen, inhibiting other people from accessing that interactive display.

before bringing the prototype to the museum. In general, literature on pervasive displays reports three main reasons why people do not interact with public displays.

6.2.1   Display Blindness

Though it seems far-fetched that visitors could miss a large interactive system, in highly stimulating environments like museums and science centers, a passerby easily may not even notice the display (Cheung et al. [2014]). For example, a group of museum visitors could walk through the interaction space without stopping or even glancing at the interactive screen. This can frequently happen in crowded spaces, or in environments in which there are many competing distractions (e.g., the other exhibits at the museum). Proxemic Interaction (Greenberg et al. [2011])—i.e., interactive devices that adapt to the user's fine-grain position within the interaction space—has been used as a strategy to entice people towards the display. For example, Hello.Wall (Prante et al. [2003]) is a wall display that shows different light patterns depending on the user's proximity to the display, while Proxemic Peddler (Wang et al. [2012]) cycles through different commercial advertisements depending on a passerby's proximity to a large screen.

6.2.2 Interaction Blindness

Other times, people notice the screen, but they do not know that they can interact with it (Huang et al. [2008]). Imagine, for example, a museum visitor looking at the data visualization on the screen, without making any attempt to control it with gestures or body movements. Sometimes, the problem is that people are not familiar with novel interaction styles. Ojala et al. [2012] observe that people may not interact with a screen because they "simply do not know that they can" and suggest placing a keyboard and mouse in front of the screen (because people are familiar with such interactive tools). This solution, however, does not fit the design of gesture-based interactions. It is also possible that this problem could be bound to disappear with time: for example, just over a decade, ago Hornecker [2008] reported that museum visitors had a difficult time using a touch-screen in a museum. Nowadays, on the contrary, basic touch screen gestures (such as pinch and zoom) have become part of our lexicon, up to the point that users (especially young children) tend to touch non-interactive screens in museum galleries (Trajkova et al. [2020a]) and their TVs at home.

6.2.3 Affordance Blindness

Even after they realize that a display is interactive, users may not understand how to use it (Coenen et al. [2017]). This is the case, for example, when a visitor tries to perform gestures and body movements to control a data visualization, but is not able to guess the correct ones to properly operate the system. In museums, this challenge is particularly problematic because visitors do not consult user manuals on how to operate a system; if the exhibit does not respond to their control actions, they will quickly leave (Cafaro et al. [2013]). Specifically, the work in Mishra and Cafaro [2018] reports that museum visitors who were not able to guess how to operate a Kinect-based data visualization (on a 65 in screen) after trying for 10 to 15 seconds all gave up, visibly frustrated that the visualization could not respond to their movements.

6.3 Providing Entry Points to the Interaction

We mentioned that the first 10–15 seconds are crucial for creating engaging HDI experiences in museums. This time is even shorter than the two minutes that visitors generally spend with traditional exhibits and artifacts (Sandifer [1997]). The question then arises: What interaction strategies can we implement to design engaging HDI systems?

The study in Mishra and Cafaro [2018] considers four strategies—although their evaluation is limited to the researchers' observations of a pool of 56 museum visitors.

FIGURE 6.2: Instrumenting the interaction space: the word "stop" denotes a hotspot on the floor, and a green yoga mat highlights the area in which users are able to control a timeline on the screen. (*Source*: Mishra and Cafaro [2018].)

6.3.1   Instrumenting the Floor

This strategy involves instrumenting the floor (within the interaction space) by adding colorful mats and writing words like "zoom in" and "timeline" to denote functionalities that people may activate or control when they step at specific locations (see Figure 6.2).

This strategy, however, does not seem to be particularly effective: people just kept looking at the screen without noticing what they were stepping on.

6.3.2   Forcing Collaboration

The screen was grayed out and displayed a message inviting people to bring a friend until two or more people entered the interaction area. In some sense, this strategy was supposed to facilitate the *honeypot effect*, a phenomenon discovered in the seminal work by Brignull and Rogers [2003]: the more people congregate around the screen, the more others tend to stop by (see Figure 6.3).

This strategy, however, does not seem to be effective: one third of the people who tried using the system just left the interaction space (rather than calling a partner). Probably, museum visitors want to immediately interact and do not like the idea of waiting for a companion before being able to use an installation.

FIGURE 6.3: Honeypot effect: People tend to congregate in larger numbers in front of screens when other people are already present (left) than when nobody is around (right).

6.3.3   Implementing Multiple Gestures to Control the Same Effect

In this case, users were able to control a timeline by jumping up and down, swaying side by side, or rocking back and forth without moving their feet. Different interactions, then, all generated the same effect on the data visualization.

This strategy seems to be effective against interaction and affordance blindness, because people have more chances to guess one of the ways to operate the system.

6.3.4   Visualizing the Visitor's Silhouette Beside the Data Visualization

Finally, introducing a representation of the user on the screen seems to increase the time that visitors spend interacting with the installation, although it could be distracting.

A basic implementation is shown in Figure 6.4: a sample application from the Microsoft Kinect SDK runs next to a custom-made data visualization and displays the silhouette of the user.

This strategy is in line with the findings from the literature on public displays. Müller et al. [2012] investigated how passersby notice the interactivity of public displays when they are represented as a mirror image, skeleton, avatar, or an abstract pattern on the screen. In that work, the interactive screen displayed a ball game. In-lab, mirror images and skeletons were more efficient than avatars and abstract patterns in attracting people towards the screen. In-situ, the mirror image performed best. Similarly, Ackad et al. [2016] compared skeletons vs. silhouettes using a public information display and found that skeletons support longer interactions than silhouettes

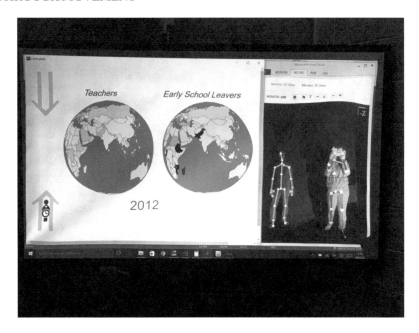

FIGURE 6.4: A user's silhouette and skeleton are displayed next to the data visualization. (*Source*: Mishra and Cafaro [2018].)

and facilitate play, while silhouettes attract more passersby and are better suited for more serious interactions.

6.4   Representing the User as Camera Overlay, Silhouette, and Skeleton

The existing literature on public displays, however, cannot be easily translated into design recommendations for HDI installations: the application scenario is just too different. The study in Trajkova et al. [2020a] thus explores how these design ideas play out when the display does not show games, reminders, or advertisements but an interactive data visualization. It reports the results of a study with 731 museum visitors and investigates how different ways of representing the user next to a data visualization impact the interaction with an HDI system.

Specifically, the user was represented either: (1) as a Skeleton; (2) as an Avatar; or (3) using a full Camera Overlay. Additionally, a Control condition was considered, in which the user was not represented at all: the display only showed a globe-based data visualization. These alternative design ideas are illustrated in Figure 6.5.

It turns out that these three different ways of representing the user impact: (1) the ability of the HDI installation to attract visitors that walk past the screen; (2) the gestures and body movements that people do in front of the installation; and (3) the time that visitors spend looking at the data.

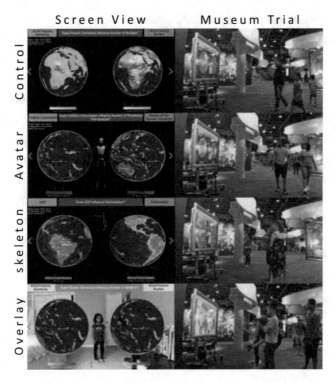

FIGURE 6.5: Four versions of the IDEA (Interactive Data and Embodied Analysis) prototype installation. (*Source*: Trajkova et al. [2020a].)

6.4.1 Ability to Attract Passersby

Findings from Trajkova et al. [2020a] indicate that the best strategies to attract more visitors towards the screen are either representing the user as an Avatar or showing the full Camera Overlay. In other words, Avatar and Camera Overlay mitigate the issues of display and interaction blindness that we discussed earlier.

In order to measure how many passersby get lured towards the screen, we can use the "capture rate" (J. Roberts et al. [2018]), i.e., the proportion of visitors who decide to interact with the system among all of those who have the opportunity to engage: they are not only near enough to the display to realize it is interactive; they also look at it to realize it is there (see Section 6.2.1). This metric is defined as:

$$CaptureRate = NumberOfUsers/NumberOfVisitors$$

In the experiment in Trajkova et al. [2020a], the capture rate increased from Skeleton (M = 0.17, SD = 0.06) to Control (M = 0.25, SD = 0.09) to Avatar (M = 0.37, SD = 0.11) to Camera Overlay (M = 0.38, SD = 0.105).

6.4.2   Influence on Gestures and Body Movements

Additionally, people seem to interact with the data visualization in different ways depending on how they are represented on the screen. In other words, by knowing how the user is represented on the screen, we can better predict the gestures and body movements that museum visitors will most likely do in front of the screen. For example, people in the Camera Overlay condition predominantly walked back and forth within the interaction space or danced in front of the screen, while most users in the Control condition opted for a pan-like gesture.

The way in which people are represented on the screen also alters the amount of time that visitors spend performing specific gestures. For example, users interacting in the Camera Overlay condition spent more time dancing ($M = 7.48, SD = .503$) than jumping ($M = 2.04, SD = .809$). In the Control condition (the user was not represented near the data visualization), people spent more time touching the display ($M = 14.72, SD = 1.42$) than dancing ($M = 5.91, SD = 1.71$).

In summary, if we are able to understand how the visualization influences the gestures and body movements that people do in front of the screen, we can then provide guidelines on how to design the interaction in a way that limits interaction and affordance blindness.

6.4.3   Effect on the Amount of Time Visitors Spend Looking at the Visualization

Finally, people spend more or less time looking at the actual data on the screen depending on how they are represented on the screen.

Museum visitors spent more time looking at the data when they were represented as an Avatar ($M = 14.78, SD = 1.13$) than when they were shown as a Skeleton ($M = 7.86, SD = .82$), using the full Camera Overlay ($M = 7.6, SD = .58$), or not visualized at all on the screen ($M = 6.98, SD = 1.23$).

6.4.4   Implications for HDI and Future Research Directions

These findings have a broader design implication for HDI: the way in which we represent the user on the screen needs to match the intended use of the system. For example, if the data visualization is designed to enable fine-grained data analysis tasks, it should represent the user in a way that promotes fine-grained interaction, such as pan-like gestures (Control). Vice-versa, organic HDI activities that focus on exploration and play may be better supported by Avatar, Skeleton or Camera Overlay representations.

Like in many research problems, however, there is still a significant amount of work to do to further dig into the specific effect that each way of representing the users has on the interaction. Along these lines, the work in Lee-Cultura et al. [2020], which focuses on one specific modality (avatars), provides prescriptive design recommendations for crafting avatars for educational games. This research study compared three different types of avatar in gesture-controlled educational games for children between 8 and 12 years old: a white hand that resembled the user's hand (low self-representation), a fictional character (moderate self-representation), and a realistic image of the player (high self-representation). Findings from this study show that

- the low self-representation avatar (white hand) minimizes fatigue when using the system and enabled longer interactions;
- the moderate self-representation avatar (character) increases the number of movements that people perform (especially when they try to explore what the character can do when they move);
- the high self-representation avatar increases engagement and arousal, which leads Lee-Cultura et al. [2020] to hypothesize that this might be the most suitable type of avatar to support learning.

Similar work could be done to investigate different types of skeleton and silhouette representations, as well as different ways to show the full camera overlay (e.g., Should we remove the background? How big should the user be compared to the data visualization?). Additionally, work that deals with different levels of realism should be aware of the uncanny valley (Cafaro et al. [2014b], Mori et al. [2012]), i.e., the possibility of creating eerie experiences for the users when the avatar looks and acts almost (but not fully) like an human. Finally, it would be interesting to use structured, visual approaches to evaluate the effect of different HDI prototypes on visitors' paths through the museum—in particular, methods from time geography as indicated in B. Shapiro and Hall [2018] and B. Shapiro et al. [2017].

6.5   Summary: HDI, Informal Learning, and Design

In this chapter, we discussed challenges and design strategies for engaging people with HDI and explained how changing the way in which the user is represented on a shared screen has an impact on the capture rate, on the type of gestures and body movements that people are willing to perform, and on the time that visitors spend engaging in organic HDI.

In Chapter 7, we will describe approaches for designing the hand gestures and body movements that people can then use to interact with HDI installations.

C H A P T E R 7

Designing Hand Gestures and Body Movements for HDI

So far in this volume we have introduced theoretical underpinnings and practical considerations for HDI in public settings, but an essential ingredient for the design of HDI remains: how to craft hand gestures and body movements that visitors can use to interact with these installations. After all, HDI exhibits are useless if nobody can figure out how to control the data visualization. This challenge is particularly relevant in informal learning settings, as museum visitors do not typically consult user manuals before exploring an installation (Cafaro et al. [2013]). Thus, we need approaches and methodologies to inform the design of these control actions.

This challenge is the focus of this chapter. First, we provide an overview of how gestures have traditionally been classified in the literature. Next, we present a few top-down guidelines for gesture design. Last, we discuss how pools of potential users can participate in the design process thanks to elicitation studies (Wobbrock [2006]).

7.1   Taxonomies of Human Gestures

Hand gestures and body movements have been studied since Ancient Roman times. Although these taxonomies do not focus on technology, they provide insights on how gestures and body movements are executed, on their intended purpose, and on how they differ from each other.

In *De Oratore*, Cicero explains that a good public speaker should master the use of bold gestures from martial and gymnastic exercises, because such gestures are as valuable as the tone of voice for delivering effective talks. In "Ars Oratoria," Quintilian describes specific examples, such as "in continuous and flowing passages a most becoming gesture is slightly to extend the arm with shoulders well thrown back and the fingers opening as the hand moves forward."[1]

These works are perhaps the first attempts to classify human gestures. As Dutsch [2002] points out, Quintilian divides all human gestures into imitative and natural; natural gestures are

---

1. Book XI, Chapter 3, line 84, as translated in the Loeb Classical Library Edition 1920, accessed online on 9/22/2020 at: https://penelope.uchicago.edu/Thayer/E/Roman/Texts/Quintilian/Institutio_Oratoria.

those that we use while we speak and could include "hand movements equivalent to adverbs, pronouns, nouns, and verbs." Additionally, Quintilian organizes gestures by body parts: head, eye and eyebrow, postures, nostrils and lips, neck, shoulder, torso and arms, hands, and feet (Kendon [2004])—an approach that is still relevant when implementing gestures and body movements for HDI in a procedural fashion.

Through history, authors and thinkers have proposed different taxonomies. In his review of gestures, Kendon explains that "different classification schemes ... reflect the different ways in which gesture has been viewed as a form of expression or communication ... Gestures have been classified according to different criteria, such as whether they are voluntary or involuntary; natural or conventional; ... how they are related to speech; their semantic domain" (Kendon [2004]).

7.1.1   McNeill Taxonomy of "Spontaneous" Gestures

McNeill [1992] provides a comprehensive psycholinguistics[2] analysis of the gestures that people "spontaneously" do when they speak. According to McNeill, "gestures are an integral part of language as much as words, phrases, and sentences" (p. 2).

In the first chapter of the book, McNeill [1992] provides a taxonomy of the spontaneous gestures that people use while speaking: they can be classified using four distinct categories.

- **Iconic.** These gestures are pictorial representations of a concrete object or event (e.g., clapping hands while talking about a car accident). Iconic gestures may also reveal the speaker's perspective and thinking process. For example, a speaker may perform a gesture in which he appears to grip something and pull it back when describing somebody bending a tree to the ground (p. 12). This gesture shows that the speaker is taking the perspective of the agent rather than trying to embody the tree.

- **Metaphoric.** These gestures represent an abstract concept (e.g., mimic a scale when talking about social justice). Similar to Iconic gestures, Metaphoric gestures are pictorial representations that we make with our hands or our body. The gesture itself depicts a metaphor for a concept, i.e., an image that the speaker feels is similar/connected to an abstract concept. McNeill explains that the gestures in this category are strictly connected with Lakoff's Conceptual Metaphor Theory[3] (Lakoff et al. [1999]). For example, a speaker may keep her hands apart as if she was representing a box when describing a cartoon on TV (McNeill [1992], p. 13). The gesture in this example is metaphorical because it represents a genre (a cartoon, the destination domain) as a bounded region of space (a CONTAINER, the source domain). Thus, according to McNeill, this type of gesture is "one of the most important ...

---

2. From the *Oxford Dictionary:* "The study of the relationships between linguistic behaviour and psychological processes, including the process of language acquisition."

3. We reviewed Lakoff and Johnson's Conceptual Metaphor Theory (CMT) in Chapter 3.

since it shows how abstract thought can be carried out in terms of representation of concrete objects and time and space" (p. x).

- **Deictic.** They are pointing gestures. Typically, people point to indicate objects while talking about them (e.g., pointing to a specific object when describing it). Pointing can, however, be abstract: for example, a person may point to the space around her when asking another person where he comes from. In the latter case, the pointing gesture does not directly refer to the space around the two people but to a different place or city (p. 18).
- **Beats.** These gestures mark specific elements of a conversation (e.g., touching the table three times while introducing three concepts). "The typical beat is a simple flick of the hand or fingers up and down or back and forth; the movement is short and quick" (p. 15).

An additional group of gestures—which, however, overlaps with the prior classifications—is called **Cohesive**: they are iconic, metaphorical, deictis, or beats gestures that are used multiple times during a conversation to provide an idea of continuity, in particular the "recurrence or continuation of a theme" (p. 16).

Thus, designers of HDI interfaces should be aware of the range of gestures and body movements that people may—voluntarily or involuntarily—perform while they talk with each other in an informal learning setting. Some of these gestures may be meaningful to control the HDI installation; others may need to be filtered out.

7.1.2   Kendon's Continuum of Human Gestures

Kendon [1988] introduced the idea that gestures are an "intermittent" activity (people do not typically accompany all their speech with gestures) and that they can be more or less structured (ranging from being like words in a language to simpler graphical representation). Additionally, when a gesture is well established within a community, it may even be used instead of words to complete a spoken sentence.

In later work, McNeill [1992] arranges gestures along a continuum that he called Kendon's Continuum of Human Gestures. Moving along the continuum means moving through these five different classes of gestures:

- **Gesticulations.** They are the most frequent type of gestures: spontaneous gestures, which accompany speech and are almost always made with arms and hands.
- **Language-Like Gestures.** They are similar to gesticulations, but they are fully integrated in a sentence and used in lieu of words—thus, they are fully integrated into the "utterance" (i.e., a piece of speech that starts and ends with a silence).
- **Pantomime**. They are sequences of gestures used to represent objects or actions; they may occur with or without speech.

TABLE 7.1: Kendon's Continuum of Human Gestures

Gesticulations	Language-Like Gestures	Pantomime	Emblems	Sign Languages
+	Speech accompanies the gesture			−
−	The gesture shows the properties of language		+	

Source: McNeill [1992].

- **Emblems.** They are highly conventionalized gestures that are easily interpreted within a similar cultural group. They are meaningful without speech, although they may also occur with speech. They follow standards: for example, the "ok" sign is made "by placing the thumb and the index fingers in contact" (McNeill [1992]). They need to be learned, and they are therefore teachable. Other common examples are the thumb-up sign and less polite gestures. Emblems usually have long historical roots, and they may even precede the languages that we speak today: for instance, the medium (middle) finger sign has been understood since Julius Caesar's time (McNeill [2006]).

- **Sign Languages.** They are the letters or words of a sign language, such as the American Sign Language. They follow grammar rules similar to those of spoken languages.

Moving from gesticulations to sign languages, the degree to which speech accompanies a gesture decreases, and the degree to which the gesture follows well-established formal rules increases (see Table 7.1).

It is worth noting that there are characteristics of human gestures that are not captured in Kendon's continuum: for example, the hands may be used as if they represented objects or to describe objects ("Object-Viewpoint"), or even as if they were the hands of some characters ("Character-Viewpoint") (see Kendon [2004], p. 105). Thus, Kendon argues that it would be better to use multiple continua to properly understand gestures—for example, each category in the continuum could actually be a continuum by itself.

7.1.3 Toward Gesture Classification Systems for Interaction Design

As observed in Wobbrock et al. [2009], McNeill's taxonomy has been designed to analyze and understand human discourses. As a result, it cannot be easily translated into design recommendations for crafting gestures and body movements that people can then use to provide input to an HDI installation.

Most likely, HDI gestures correspond to the Emblems in Kendon's Continuum of Human Gestures: they are well defined, they can be learned, and they are meaningful with or without speech. There is an additional factor that designers of HDI installations should keep in mind. Specifically, the research on emblems focuses on the creation of lists of emblems that are common within specific communities or ethnic groups (Kendon [1992]): for example, people in Naples, Italy, use a particularly wide vocabulary of gestures (Kendon [2004]). Because these catalogs focus on well-defined communities, they are typically difficult to generalize for the design of embodied interactions (unless the interactive system is specifically created, for instance, for people who live in Naples). This does not mean, however, that gesture classification systems are useless for the design of HDI. There are, for example, some emblems that are almost universally recognized, such as the thumb's-up sign.

7.2   Approaches to Gesture Design

In this section, we review two groups of design guidelines for crafting gestures: one based on extending the gesture taxonomies that we reviewed above, the other on expanding design guidelines that were originally conceived for traditional windows-icons-menus-pointer (WIMP) interfaces.

7.2.1   Extending Gesture Taxonomies

Gesture taxonomies provide the theoretical foundation for HCI-oriented taxonomies, like the one in Karam et al. [2005]. In particular, the work in Karam et al. reports an extensive literature review of gesture-based interactions and proposes to classify gestures according to four characteristics:

- **Application domain**, for example, if the gestures are meant to be used with desktop applications, virtual/augmented reality, games.
- **Enabling technology**, such as keyboard, mouse, tangible objects, touch surfaces.
- **Gesture style**, which is described in a similar way as McNeill's taxonomy but using different categories, precisely, deietic, gesticulations, semaphors (used to provide simple signs—like start/stop—to computer interfaces), sign language.
- **System responses**, like video or audio output.

The reader should keep in mind, however, that the categories and labels in Karam et al. [2005] come from a literature review and not directly from linguistic analysis. So, for example, the word "gesticulations" is used in this taxonomy because it has been used in the literature, even though Kendon simply used it in earlier works as a synonym of "gestures" and later moved away from it (Kendon [2004]).

The main takeaway from this work is the notion that application domain, enabling technology, and feedback should all play a role in the design of gestures and body movements for HDI. For instance, when designing the interaction with a smaller screen like a smartphone, we should probably avoid full-body movements that require users to stand far away from the device camera (because people typically hold their phone in their hand), while when crafting gestures and body movements for large displays, we can include more dynamic, full-body interactions and avoid smaller gestures that will be difficult to read for whole-body-tracking camera systems.

7.2.2   Extending WIMP Guidelines

According to Norman [2010], there is a desperate need for design guidelines that could help the process of crafting and implementing gestures and body movements:

> When the Nintendo Wii introduced its bowling game, the natural interface was to swing the arm as if holding a bowling ball, and then, when the player's arm reached the point where the ball was to be released, to release the pressure on the hand-held controller's switch. Releasing the pressure on the switch was analogous to releasing the ball from the hand. Alas, in the heat of the game, players would also release their hand pressure on the controller which would fly through the air, sometimes with enough force to hit and break the television screen.

The solution to this problem, according to Norman and Nielsen [2010], is to define design guidelines that build upon established principles used in WIMP interfaces:

- **Visibility** (also called Perceived Affordances/Signifiers). The system should communicate which gestures are implemented and the application scenarios in which they apply.
- **Feedback**. Users should be made aware of the effect of their actions to avoid situations like the one that occurs when people press the back button on their phone and, as a result, exit their app (rather than going to a previous page).
- **Consistency**. Similar gestures should trigger the same operation in different applications.
- **Non-Destructive Operations**. There should be a way to undo an action.
- **Discoverability**. Users should be able to figure out which gestures they can use (so that they do not need to memorize specific commands, like in the case of shell-based interfaces).
- **Scalability**. Gestures should work with different screen sizes, from small screens to large displays.
- **Reliability**. The interface should do its best to recognize the users' gestures, even when they are slightly different from the way in which they were originally envisioned by the system designers—otherwise, users get confused on what their actions do, creating a sense of randomness.

In general, applying WIMP design principles to full-body interaction is a challenging task, because gestural interfaces do not force users to be seated at a desktop and are not constrained by the layout and the number of keys/buttons of a typical keyboard or mouse; rather, users can potentially do anything that a camera is able to track within its field of view. In a broad sense, however, they are a useful tool when crafting and comparing methodologies and approaches for designing gestures and body movements. For example, we will see later in this chapter how the Framed Guessability approach in Cafaro et al. [2018] focuses on improving the "discoverability" of controlling actions.

```
   Welcome. For the next hour or so you will be working with an
experimental system designed to process computerized mail. This system is
different than other systems since it has been specifically designed to
accomodate to a wide variety of users. You should feel free to experiment
with the system and to give it commands which seem natural and logical to
use.

   Today we want you to use the system to accomplish the following tasks:

   1) have the computer tell you the time
   2) get rid of any memo which is about morale.
   3) look at the contents of each of the memos from Dingee.
   4) one of the Dingee memos you've looked at contains information
      about the women's support group, get rid of that message.
   5) on the attached page there is a handwritten memo, using the
      computer, have Dennis Wixon receive it.
   6) see that Mike Good gets the message about the keyboard
      study from Crowling.
   7) check the mail from John Whiteside and see that Burrows gets
      the one that describes the transfer command.
   8) it turns out that the Dingee memo you got rid of is
      needed after all, so go back and get it.
```

NOTE: The above is an *exact* replica of the task given for the UDI experiment.

FIGURE 7.1: Tasks used in elicitation experiment. (*Source*: Good et al. [1984], p. 1033.)

7.3   The State-of-The-Art of Gesture Design: Elicitation Studies

Elicitation studies are a bottom-up design approach in which gestures and body movements are grounded on recommendations made by pools of potential users. This technique is particularly useful when designing for human-data interaction, because there are not many a-priori guidelines on how to craft gestures and body movements that people can use to interact with data visualizations.

7.3.1   An Early Example of Elicitation Studies

Good et al. [1984] describe a design approach that is perhaps the first attempt to use elicitation studies for designing interactive systems. The overarching idea was that computer systems and interfaces should not be designed "with the assumption that the user must adapt to the system, that users must be trained and their behavior altered to fit a given interface." On the contrary, computer systems should be crafted in a way that resembles what novice users tend to do when they interact with them. In the experiment reported in Good et al., 67 novice computer users were asked to complete eight tasks related to electronic mail (a novelty at that time) using a textual interface (see Figure 7.1).

Users were asked to type commands of their choice into a terminal screen. Using a Wizard of Oz approach, if the system was not able to recognize the command, a hidden human operator translated it into machine language and executed it, so that the system always looked reactive to the user. All the commands were recorded in a log and used to develop a new, more intuitive interface

FIGURE 7.2: The Palm TX PDA (2005). (*Source*: Museo8bits, CC BY-SA 3.0, https://commons. wikimedia.org/w/index.php?curid=944858.)

that was able to recognize 86% of the 1070 spontaneous commands that were recorded during this study (Good et al. [1984]). More recently, a similar type of elicitation has been used to identify natural language commands for a visualization interface. Subjects were shown images of various charts and asked to describe them in natural language in service of the creation of a toolkit to support natural language-driven data visualization (Narechania et al. [2020]).

7.3.2 Wobbrock's Guessability Studies

In late 1990s and early 2000s, well before the age of smartphones, a new type of device started to gain more and more popularity: the Personal Digital Assistant (PDA; see Figure 7.2). A significant challenge for these interfaces was how to recognize hand characters that users input with a stylus. Typically, people had to write using the artificial rules of single-stroke recognition systems, e.g., Unistrokes or Graffiti (Castellucci and MacKenzie [2008]).

This is the technological landscape in which Wobbrock et al. [2005] introduced the idea of Guessability Studies. The concept of "guessability" is defined as:

> that quality of symbols which allows a user to access intended referents via those symbols, despite a lack of knowledge of those symbols. (Wobbrock et al. [2005])

Wobbrock et al. [2005] argue that it is crucial to assure high guessability when people need to use stylus strokes to input ASCII characters on small, handheld devices. Thus, they introduce an elicitation methodology for maximizing the guessability of symbolic inputs. They also highlight how the concept of guessability is different from the idea of immediate usability (MacKenzie and Zhang [1997]): rather than evaluating the user's experience after a brief learning period (for example, by letting participants familiarize themselves with a new system before they actually have to use it), guessability involves no prior learning.

Specifically, in Guessability Studies, groups of potential users are invited, one at a time, into a lab room. Participants are only given basic, essential information (e.g., that the system required unistroke symbols rather than multi-stroke) and are not provided with any examples. Each person is then

1. exposed to a "referent" (i.e., a letter); and
2. asked to propose a "symbol" (i.e., a stroke) for that referent.

For each participant, the procedure is repeated for all the "referents" for which the designers needed to identify suitable "symbols."

At the end of the study, each referent is assigned the symbol (in this case, a stroke) that was recommended by the biggest number of participants. In case of conflicts (different users may recommend the same symbol for different referents), the symbol is assigned to the referent for which it had been recommended the most. This process allows designers to build a user-defined set of symbols.

7.3.2.1 The Agreement Metric

In order to evaluate the user-defined set of symbols, Wobbrock et al. [2005] introduced a custom-made metric called "Agreement." Mathematically, the Agreement A among the symbols that were recommended by participants is defined as the percentage

$$A = \frac{\sum_{r \in R} \sum_{P_i \subseteq P_r} \left(\frac{P_i}{P_r}\right)^2}{|R|} * 100,$$

where r is a referent in the set of all referents R, P_r is the set of proposed symbols for referent r, and P_i is the subset of identical symbols from P_r (Wobbrock et al. [2005]).

This measure is designed to reach 100% when all participants recommend the same symbols, and tend to 0% when they all come up with different ideas. Thus, the Agreement measure is used as a proxy for the Guessability property: the underpinning idea is that, if participants tend to agree a lot during the lab study, then the actual users of the system will also easily guess how to interact with it after the system is deployed in the intended context of use. We want to warn the reader, however, that this assumption has not been extensively validated *in-situ*.

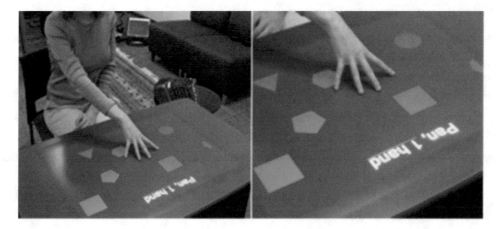

FIGURE 7.3: A participant in a Guessability Study recommends a gesture to control a functionality of the system ("panning effect"). (*Source*: Wobbrock et al. [2009].)

7.3.3   Guessability Studies for Gesture Design

Guessability studies have not been confined to small PDA screens for long. Wobbrock et al. [2009] extended this methodology to the problem of defining gestures for surface computing (i.e., table-size touchscreens) (see Figure 7.3). In this case, the "referents" were functionalities of the table-size screen, such as "selecting" or "moving" a shape, presented as animations; the "symbols" were the gestures (e.g., tap, drag) that participants recommended. The methodology remained unaltered.

Since its extension to surface computing, guessability studies have been applied to a broad range of gesture design domains. For example, Ruiz et al. [2011] used a guessability study to construct a set of motion gestures for smartphones. Examples from the resulting gesture set are: placing the phone to the ear to answer a phone call, rotating the phone so that the screen is away to hang up, shaking the phone to go to the home page, etc.

For the curious reader, the work in Villarreal-Narvaez et al. [2020] provides an extensive review of the ways in which guessability studies have been used in different application domains, with slightly different approaches.

7.3.4   Considerations for Applying Guessability Studies to Human-Data Interaction

Although elicitation studies arguably represent the state-of-the-art when designing gestures and body movements for human-data interaction, they raise few considerations that designers of these systems should keep in mind. Even if, at a first glance, it may look like elicitation studies are based exclusively on the users' recommendations, designers still play a fundamental role when crafting control actions for HDI using this methodology.

7.3.4.1 Identifying Classes of Similar Gestures

A first challenge is that different people may showcase gestures in slightly different ways. For example, a swipe gesture may be performed moving hands at different speeds or keeping the hand at different starting positions when compared with the torso.

The work in Piumsomboon et al. [2013], for example, used guessability studies to devise gestures for augmented reality. Participants were prompted with a 3D animation of a task and then asked to recommend a gesture to perform the task. Interestingly, this is perhaps the first work to problematize the process of deciding when similar gestures recommended by different participants can be considered the same gesture. Participants generated a total of 800 gestures, which were grouped into 320 unique gestures. Two gestures were deemed similar enough if their path was identical or had a consistent directionality.

Thus, when designing for human-data interaction, we need to remember that the same mid-air gesture or body movement may be executed in slightly different ways by different people. Indeed, designers have a significant role in this context: during the elicitation, they need to deem when two gestures are the same; during the implementation, they need to craft control actions in ways that are flexible enough to be used by a variety of people.

An alternative approach could be to use the video recordings from the elicitation studies to train gesture recognition systems based on machine learning by demonstration (Trajkova et al. [2020a]). For example, the SDK for the older version of the Microsoft Kinect included a tool, called Visual Gesture Builder, that could be particularly useful in this regard (more modern approaches may involve using Microsoft Azure ML or PowerBI, or similar tools, to train neural networks or classifiers).

7.3.4.2 Balancing Legacy Biases

Legacy biases also deserve consideration when designing novel interactions for HDI. In particular, Morris et al. [2014] observes that a "potential pitfall" of elicitation studies is that people may be already familiar with older technologies and interfaces (e.g., WIMP). This brings participants to elicitation studies to recommend gestures that are based on old interaction paradigms (e.g., a mouse click). According to Morris et al., legacy bias "limits the potential of user-elicitation methodologies for producing interactions that take full advantage of the possibilities and requirements of emerging application domains, form factors, and sensing capabilities."

There is a tradeoff, however. Morris et al. [2014] observe that as legacy biases "draw upon culturally shared metaphors, participants tend to propose similar legacy-inspired interactions, resulting in high agreement scores in elicitation studies." A different opinion piece, Köpsel and Bubalo [2015], points out that legacy biases can actually help us to create good gestures, because they allow designers to build upon interactions that groups of people are already familiar with.

Designers of gestures and body movements for HDI should then be aware of the presence of legacy biases when they conduct elicitation studies and assure the delicate balance between novel control actions and others based on legacy approaches (that can reassure the user and provide entry points to the interaction). As observed in S. Williams et al. [2020], additionally, the number of elicited legacy biases may depend on the referent: for example, "pushing a button" generates more legacy biases than "moving forward" in a virtual reality space.

7.3.4.3 Eliciting Multiple Gestures

In the original formulation of guessability studies (Wobbrock et al. [2005]), participants are asked to recommend only one symbol/gesture per referent/system functionality. What happens if a participant recommends multiple gestures for one functionality? According to Morris [2012], Wobbrock's agreement metric cannot be used anymore.

In particular, Web on the Wall (Morris [2012]) used a Microsoft Kinect to control a web browser on a TV screen without a remote. Twenty-five participants were interviewed, who suggested gestures and speech interaction for 15 web browser functionalities, such as performing a search, going back, going forward, opening a link in a separate tab. Participants were allowed to recommend multiple gestures for the same referent. In such an experimental scenario, the notion of agreement should be substituted with two different metrics: the "max-consensus" score and the "consensus-distinct" ratio. The max-consensus score is the percentage of participants who suggested the most popular gesture for each referent. The consensus-distinct ratio is the fraction of distinct interactions proposed for a referent. Also, a "consensus threshold" should be met: interactions are not significant if they are not recommended by at least two participants.

This issue is particularly relevant when designing for HDI because supporting multiple gestures and body movements to execute the same system functionality is typically a good strategy for providing entry points to the interaction (we discussed this idea in Chapter 5). Thus, the traditional Agreement metric may be ill-suited for this context.

7.3.4.4 Considering Alternative Metrics

Additionally, the work in Vatavu and Wobbrock [2015] introduces two alternative metrics that can be useful to evaluate a user-generated set of gestures and body movements. Specifically:

- *Disagreement rate* between participants, defined "for a given referent r as the number of pairs of participants that are in disagreement divided by the total number of pairs of participants that could be in disagreement";
- *Coagreement rate* of two referents r1 and r2, defined as "the number of pairs of participants that are in agreement for both r1 and r2 divided by the total number of pairs of participants

that could have been in agreement." For example, this metric can be helpful to detect multiple system functionalities for which most participants tend to agree on specific gestures or body movements (e.g., if most people recommend walking closer to the screen to zoom in *and* walking away from the screen to zoom out).

7.3.4.5   Extending Guessability Studies beyond the Lab

Guessability studies are executed in a lab setting with a relatively small number of participants. As observed in Ali et al. [2019], however, there is a lack of a formal evaluation on the difference between the participants' agreement in-lab and what actually happens in-situ. Two exceptions are: (1) an early evaluation of the *accuracy* of the alphabet that was created in Wobbrock et al. [2005], which is reported to be more accurate than Graffiti for able-bodied and motor-impaired users (Wobbrock [2006]), and (2) a study on the *memorability* of gestures that showed that, after learning how to operate a system with 22 gestures, 18 participants were able to recall user-generated gestures better than those defined by designers on the following day (Nacenta et al. [2013]).

As a workaround on this problem, Ali et al. [2019] propose to conduct distributed elicitation studies via crowdsourcing. In particular, Ali et al. present Crowdlicit, a web-based interface designed to broaden participation to elicitation studies by allowing a larger pool of users to participate in the elicitation procedure online (78 participants from Amazon's Mechanical Turk). Additionally, the Crowdlicit system allows designers to employ a different group of online users to validate the user-generated gesture set. This addition to the Guessability methodology is called the *identification study* and is described as "the reverse of an elicitation study": participants are asked to imagine open-ended system functionalities proposals for all the symbols/gestures that were previously elicited. To evaluate the identification study, Ali et al. uses metrics similar to those traditionally adopted in elicitation studies (referent agreement).

Designers of HDI systems should be aware of this intrinsic limitation of elicitation studies (being conducted in a lab setting) and develop strategies to assure that what looks good in-lab actually works in-situ. The crowdsourcing approach offers an interesting opportunity. Alternatively, when it comes to public spaces and informal learning settings, we may need to re-think the elicitation methodology so that it can be applied in-situ, for example, by video recording the gestures and body movements that visitors spontaneously do in front of big screens (Trajkova et al. [2020a]).

7.3.5   Framed Guessability

Even with all the previous considerations, however, guessability studies suffer from an additional shortcoming: they may produce gestures that are intuitive individually but are pretty much unrelated. For example, Vatavu and Zaiti [2014] conducted a guessability study to identify gestures to control smart TVs. At the end, the user-generated gesture set included control actions that did not have much in common, such as "moving hand to the left" to go to the next channel and "drawing a letter M" to open a menu (Vatavu and Zaiti [2014]). As a result, it may be difficult for the user to discover one gesture after the other.

In order to circumvent this problem, the Framed Guessability approach (Cafaro et al. [2018]) posits that participants should be primed with a frame[4] before the elicitation study begins. The underpinning idea is that frames can structure and limit the range of possible gestures and body movements that people then recommend during the elicitation process (Cafaro et al. [2014a]): for example, one would not expect to dive into a pool while on an airplane. Results indicated that the gestures generated using Framed Guessability were overall easier for museum visitors to discover than those created with a traditional guessability approach, even when all the references to the priming frame were removed (see Cafaro et al. [2018]). In the following, we dig more into this design methodology.

7.3.5.1   Procedure: Priming and Elicitation

Similarly to guessability studies (Wobbrock et al. [2005]), participants are interviewed one at a time in a lab setting. Framed Guessability, however, introduces a preliminary priming phase that is administered before the elicitation begins. Indeed, participants go through two complementary phases described in Cafaro et al. [2018]:

1. **Priming phase.** In the first phase of Framed Guessability, a participant is primed with a frame (e.g., "Eating a steak at a restaurant"). Priming is achieved with a combination of
    *   *visual priming*, i.e., pictures of the frame are displayed on a screen as a slideshow when the participant enters the lab;
    *   *written task*, in which the participant is asked to write things that she would do in the scenario on a worksheet (see Figure 7.4 for an example);
    *   *embodied priming*, in which the participant is asked to re-enact what she wrote in the written task.

2. **Elicitation phase.** The second phase of Framed Guessability immediately follows the priming phase and is structured as a traditional guessability study. The same participant is exposed to a sequence of effects (i.e., functionalities of the system) and asked to recommend one gesture or body movement to "control" that effect.

7.3.5.2   Effect on the Discoverability of Gestures and Body Movements

Eighty-nine people participated to the Framed Guessability activities (in-lab) in Cafaro et al. [2018]. Participants were split in three groups: FUNHOUSE (primed with being at a funhouse in front of

---

4. The reader may want to review the overview of embodied schemata, metaphors, and frames in Chapter 3.

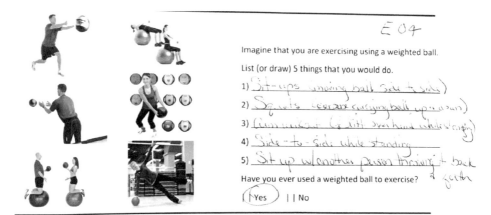

FIGURE 7.4: Worksheet for the priming task used during the priming phase of the Framed Guessability study described in Cafaro et al. [2018].

a distorted mirror); GYM (primed with being at a gym with a weighted ball); and CONTROL (no priming). This resulted in three different sets of gestures and body movements.

The Framed Guessability approach was then evaluated in a quasi experimental study with 138 museum visitors at a science museum (the New York Hall of Science in Queens, NY; see Cafaro et al. [2018]). During the in-situ evaluation, museum visitors interacted with a version of the HDI installation. They had no idea of what the frame was: there was absolutely no reference to the priming frame at the museum. Using a Wizard of Oz approach, a moderator activated an effect to participants as soon as one participant was able to guess one of the control actions in that experimental condition (FUNHOUSE, GYM, CONTROL). The number of gestures and body movements that participants were able to guess in each experimental condition is illustrated in Figure 7.5 and shows how the gestures generated with Framed Guessability were overall more discoverable than those created with traditional guessability studies. The statistical analysis revealed that the number of control actions that people discovered was significantly higher in the FUNHOUSE (M = 3.07, SD = 1.04) than in the CONTROL condition (M = 2.04, SD = 0.98). These results are interpreted as being due to the fact that the gestures and body movements generated with Framed Guessability are all interconnected: "if one schema is activated in the mind of a person, its activation can spread to other schema in an associative pattern formed via real-world experiences" (Cafaro et al. [2018]).

7.3.5.3   The Role of the Frame

As we mentioned, at the end of the in-lab Framed Guessability study, different priming frames resulted in different sets of gestures and body movements, i.e., participants recommended different control actions (see Figure 7.6).

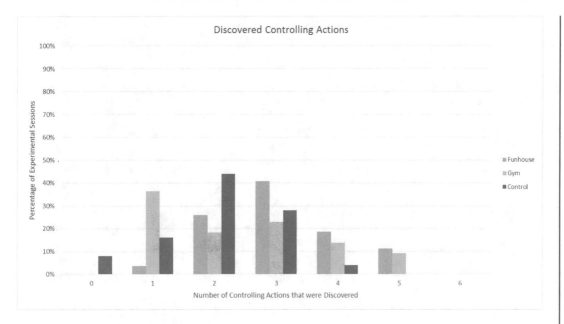

FIGURE 7.5: Percent of the 74 in-situ experimental sessions in which the number of actions guessed by museum visitors was either zero, one, two, three, four, five, or six. (*Source*: Cafaro et al. [2018].)

In particular, GYM was more biased towards full-body movement and FUNHOUSE towards arm gestures and static poses. When using the Framed Guessability approach, thus, designers of HDI installations are tasked with an important decision: selecting a priming frame. Although there is not a one-size-fits-all kind of approach, the work in Cafaro et al. [2018] discusses two broad design recommendations:

- *Discrete vs. continuous gestures.* Frames should be chosen in a way that matches whether the HDI installations should rely on discrete triggers (e.g., a pose or a gesture from the American Sign Language) or continuous input (e.g., pinch-and-zoom).
- *Social space.* The frame should align with the social norms for body movement in the space where the system will be deployed (some body gestures could be ill-suited for public spaces) and with the target population (e.g., when designing for older populations, we may want to avoid dynamic movements like jumping).

Additionally, Ali et al. [2021] describe another promising approach for generating priming frames: selecting them from popular culture. Although the work by Ali et al. does not directly mention Framed Guessability, the methodological approach is similar: participants in an elicitation study are exposed to "a montage of sci-fi films depicting characters interacting with technologies

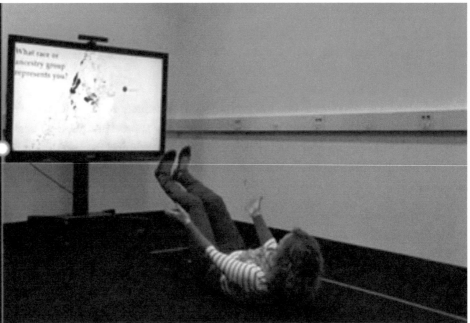

FIGURE 7.6: Elicitation portion of the Framed Guessability study described in Cafaro et al. [2018]. Two users recommend gestures and body movements in the (top) FUNHOUSE and (bottom) WEIGHTED BALL experimental conditions.

using gestures" (Ali et al. [2021]). Interestingly, the gestures produced by participants with sci-fi priming turned out to have better learnability and memorability than those generated without priming.

7.4   Summary: Gesture Classifications and Gesture Design

In this chapter, we reviewed taxonomies of gestures and top-down approaches to gesture design and discussed how these theoretical frameworks can provide guidelines for the initial stages of gesture design. We then provided an overview of elicitation studies that represent the state-of-the-art in gesture design and can be used for designing gestures and body movements to control HDI data visualizations. We concluded with a detailed overview of Framed Guessability, an elicitation-based approach that attempts to maximize the discoverability of control actions.

In the next chapter, we will show how the design of the interaction with an HDI installation impacts the way in which people interpret and discuss the data on the screen.

CHAPTER 8

Embodiment and Sensemaking

In this chapter, we present two case studies that show how embodied human-data interaction has potential to facilitate data sensemaking. We first describe how implementing embodied interaction in a multi-user census data map museum exhibit, CoCensus, supported *perspective taking* by visitors as they discussed the data and how these perspectives were related to sensemaking and learning talk. We then discuss how embodiment may support people's interpretation of causal relationships in data. Finally, we discuss two methodological approaches that can structure the task of assessing the interaction with HDI installations.

8.1   CoCensus: Embodiment to Foster Perspective Taking

We use visualizations to see and think with data. The perspective we take with respect to this data provides the lens through which we interpret it. We explored the relationship between perspective taking, embodiment, and visualization in a multi-year design-based research project called CoCensus (Figure 8.1).

8.1.1   Collaborative Exploration of Census Data

The CoCensus exhibit was developed and tested in two museum spaces—a small local history museum in Chicago, IL, and an interactive science center in Queens, NY—to investigate ways of engaging learners with the familiar yet abstract dataset of the United States decennial census and American Community Survey (www.census.gov). Unlike more formal engagements with census data one would expect in classroom interactions, CoCensus did not aim to convey a set narrative or outcome. Instead, the exhibit was designed to help individuals see themselves, or at least a reflection of themselves as defined by the census, in the data in order to enable them to compare their data with that of their companions and across time; to identify trends and hypothesize about the causes of those trends using outside knowledge; to relate the data to their lived experience in the geographic area; and to question the data itself and what it does and does not reflect about them and their identities. A productive learning interaction with this exhibit did not require acquisition of particular facts about the census or city demographics; rather, it involved substantive discussion among companions about facets of the dataset they found relevant and interesting to their own lives and experiences. For this reason the exhibit was designed to be a multi-user system, so two or more

FIGURE 8.1: CoCensus was implemented in two unique museum spaces. At the Jane Addams Hull House Museum (left), CoCensus presented ancestry and heritage data in Chicago as a modern-day complement to the Hull House maps and papers exhibition on display. The New York Hall of Science implementation (right) featured multiple census datasets including household size, house type, industry, and heritage to allow museum visitors of all ages to explore multiple facets of New York's demographics.

visitors viewing their own datasets could work together to make sense of the data by comparing and contrasting datasets with each other (see Figure 8.1).

The exhibit (described fully in J. Roberts and Lyons [2017b, 2020]) aimed to foster connections to the data by allowing visitors to self-select datasets representing them, displaying their selections as scaled centroids ("bubbles") over the local geography, and providing control to visitors to manipulate the representation via full-body movements. We anticipated that providing interaction with the data would support visitors' agency in exploring the content, and we wanted to allow the opportunity for different representations of data (e.g., different aggregation levels) to demonstrate how the same data can seemingly say different things depending on how it is represented.

During the course of this multi-year project, we iterated and tested multiple designs for the data display (J. Roberts et al. [2012, 2015]), full-body control movements (Cafaro et al. [2013, 2014a, 2014b, 2018]), and interaction area layouts (Roberts et al. [2014]) to investigate the impact of these changes on visitors' experiences. We were particularly looking for design iterations that supported the data decoding behaviors identified in Chapter 5: translating from visual to text, interpreting across multiple data points, and extrapolating beyond what is presented (Friel et al. [2001]). We measured these behaviors through analysis of "learning talk" (Allen [2003]). It is through this analysis that we identified an unexpected phenomenon: shifts in visitor perspective taking.

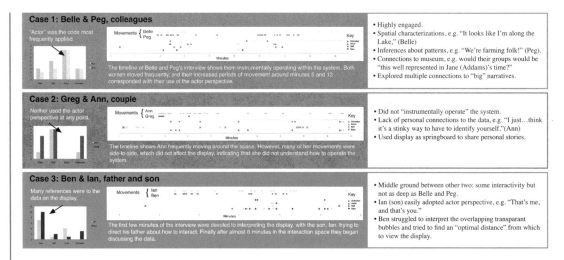

FIGURE 8.2: J. Roberts et al. [2013b] explore the relationship between movement, data interpretation talk, and perspective taking for three unique visitor pairs.

8.1.2   Early Emergence of Perspective Taking in CoCensus

When we began prototype testing with an interactive version of CoCensus using distributed, full-body visitor control, we found some visitors spontaneously used first-person pronouns when referring to the data on display (J. Roberts et al. [2013b]). Whereas in pilot testing with non-interactive mock-ups, visitors investigating heritage data had referred to the data from a third-person perspective (e.g., "The Germans are all over the North Side"). Once they were individually controlling their selected datasets through physical movements, some spoke from the perspective of someone in the map, for example, "I'm along the Lake." Temporal analysis of these sessions, during which pairs of visitors engaged in semi-structured interviews with the researchers while interacting with the display, revealed that the use of this *actor perspective* (Brunyé et al. [2009]) occurred in conjunction with visitors' body-based control movements within the interaction area. Moreover, in this small initial study, the pair of visitors demonstrating the highest frequency of actor perspective taking (APT) were the most highly engaged in the data interpretation, making inferences about populations and posing questions about relationships between their datasets over time (J. Roberts et al. [2013b]).

This linguistic shift was particularly exciting because prior work has shown potential learning benefits related to perspective taking and perspective shifting. A first-person perspective has been shown to augment performance in procedural tasks (Lindgren [2012], Lozano et al. [2006]) and during collaborative dialogue (Filipi and Wales [2004]), and it has been linked to expert decoding of data representations. For example, Ochs, Gonzales, and Jacoby [1996] analyzed physicists' conversations as they interpreted experimental results during lab meetings, finding that these experts used first-person pronouns when describing particle behaviors, such as, "When I come down, I'm in the domain state." The researchers determined the blending of identities "created through gesture, graphic representation, and talk, appears to be a valuable discursive and psychological resource as scientists work through their interpretations and come to consensus regarding research findings."

Enyedy et al. [2013] studied perspective taking in classroom science learning, positing that by looking at a problem or situation from a particular viewpoint, such as by envisioning oneself as a first-person "actor" in a scenario, learners can potentially draw upon a unique set of resources for reasoning. For example, when children envision themselves as a ball moving across a surface, they can reason about the velocity and forces of friction of the ball in a different way than they would by just watching a ball. They are able to transform their physical bodies into "components in the microworld that structure students' inferences" (Enyedy et al. [2013]). Similar use of body movements in space was used by Lindgren and Moshell [2011] in Meteor, which allowed learners to "be" asteroids traveling through space in order to reason about gravitational forces (Figure 3.8). The spontaneous appearance of this kind of talk in our interactive sessions led us to investigate whether these same principles of embodiment and perspective taking could support reasoning about mapped abstract data in CoCensus.

8.1.3   Encouraging Perspective Taking through Interaction Design

Effects of variations in the interaction design of CoCensus were explored in a December 2013 study examining affordances of two competing designs for the timeline control (J. Roberts et al. [2014]). That study analyzed dialogues between pairs of users interacting naturalistically (i.e., in an unstructured session, without any mediation by a researcher) with one of two versions of the timeline floor control. In the horizontal (H) configuration, small timeline "buttons" were placed parallel to the display, with the past (1990) on the left and more current data (2010) on the right, in alignment with standard graphical conventions where time is often represented on the horizontal axis (see Figure 8.3, left). Two visitors could participate simultaneously, but a control action from one visitor (to change the decade or category) would change both visitors' datasets. This design—left to right timeline representations and mutually exclusive timeline control—was meant to be externally consistent with common timeline representations with which visitors would be familiar. The alternative vertical (V) configuration (Figure 8.3, right) afforded separate simultaneous control of individual data sets, where stepping back (away from the display) moved back in time and stepping toward the display moved forward in time (to 2010), in an ego-moving metaphor expected to help visitors feel personally connected to the data.

Analysis of conversations from 28 participants in 14 sessions showed that the floor control configuration did impact learning talk, with the ego-moving V condition supporting more productive talk than the H condition (J. Roberts et al. [2014]). Importantly, participants in the V condition

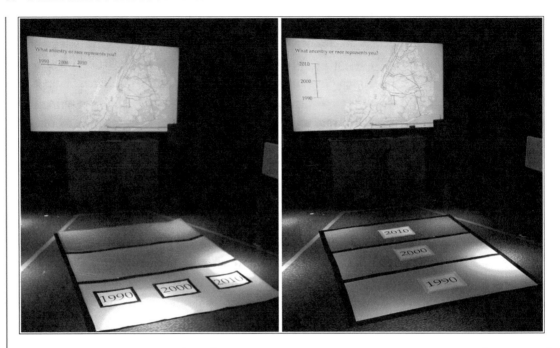

FIGURE 8.3: J. Roberts et al. [2014] use a between-subjects design to compare two configurations of timeline control by analyzing the learning talk of visitors during naturalistic interactions.

also produced significantly more statements in the actor perspective, suggesting a need for a deeper analysis of the connection between Actor Perspective Taking (APT) and learning talk.

8.1.4   Perspective Taking and Learning Talk

The initial linkages between perspective taking, specifically first-person actor perspective taking APT, and productive learning talk were encouraging but so far hardly more than anecdotal. A larger, more robust study was needed to tease out the role of APT in mediating interactions, and this study took the form of a 2 × 2 design of the control input (J. Roberts and Lyons [2017b]).

We had premised the exhibit on the idea that full-body control was crucial to supporting the desired data engagement and interactions, based on positive results from early iterations and a significant tradition of similar interactive museum exhibit designs (as discussed in Chapter 4). However, we had not yet seen empirical proof of the effectiveness of this design choice. Was full-body control really necessary, or would smaller embodied movements suffice? Did all users need individual control, or would a single "driver" controlling the visualization still lead to productive sessions? We recorded 119 sessions of visitor groups in each of four conditions in the 2 × 2 study design (Figure 8.4) and analyzed their learning talk.

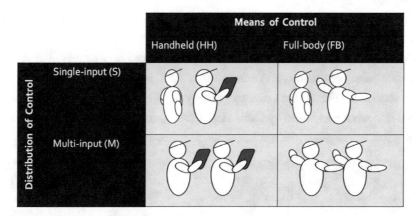

FIGURE 8.4: To elucidate the importance of the size of gestures (the means of control), we compared full-body movement conditions with handheld tablets as control devices. The two designs used congruent movements but on different scales. We also tested the distribution of control between a single-input global controller versus individual controllers for each user. See J. Roberts and Lyons [2017b] for full details.

Learning talk analysis was conducted using a novel scoring method called Scoring Qualitative Informal Learning Dialogue, or SQuILD. This complicated method (described fully in J. Roberts and Lyons [2017a, 2017b]) used simultaneous and magnitude coding of talk, meaning that snippets of dialogue were assigned multiple codes, such as "instantiating" (saying aloud the name of something on the map) or "integrating" (combining two pieces of information; e.g., making a comparison), based on a scheme developed from prior literature. These codes were given a numerical weight (one, two, or three points each) based on their relevance to the exhibit's learning goals. All values of a session's codes were tallied to create a content score for each session that could be used for comparisons across conditions to determine the "winning" design. Though developed to facilitate A/B testing of interaction design conditions, this scoring system also gave the opportunity to study spontaneous perspective taking more closely and investigate whether the correlations between APT and data talk found in the earlier studies still held up.

There was no difference among the four conditions in whether or not visitors adopted first-person actor perspectives during their sessions, with 54 of the 119 sessions including at least one APT statement. Comparing those sessions to those without APT showed that sessions in which APT was used at least once had a significantly higher total number of learning talk codes applied (M = 58.22, SD = 28.78) than sessions where no APT was used (M = 44.35, SD = 27.46), $t(117) = -2.684$, $p = .008$, and APT sessions had a higher average content score (M = 78.46, SD = 43.76) than sessions where no APT was used (M = 61.34, SD = 39.65), $t(117) = -2.238$, $p = .027$.

These findings confirmed the positive relationship between APT and learning talk seen in pilot data (J. Roberts and Lyons [2020]).

In J. Roberts and Lyons [2020] we took a deeper look into how people used APT as a mediating tool in their joint sensemaking. We found that APT was sometimes used for orientation to the map, with statements like, "That's me, and that's you." Other instances suggested a blending of identities similar to what was found by Ochs et al. [1996], with some users adopting a first-person perspective as a sort of *role play*, which we describe as follows:

> *Role-play* as identified here is a special function in which visitors are more fully adopting the subjective perspective of a character in the map, making statements like, "This is where I would work," and "I'd live downtown," and even more evaluative and extrapolative statements like, "Oh, so you're saying I'd live in like the cool area," and "I was always important." Visitors employing this kind of APT use are not just identifying quantitative patterns (e.g. where a population is highly concentrated) but are thinking more deeply about what it means to be a person with certain demographic characteristics in a certain geographical place. Role-play indeterminate constructions therefore blend the visitor with her data, thus helping to humanize the census data. (J. Roberts and Lyons [2020])

Alternatively, some visitors demonstrated a different kind of blend, *projection*:

> Some visitors used APT instead to speak for a group of people, projecting themselves onto the whole dataset. *Projection* APT statements were comments such as, "You guys dominate," "There's a lot of me," and, "You're really spread out." In each case, "I" and "you" become referents not to any one person but to the entire dataset, allowing the visitor to make more abstract observations and inferences. (J. Roberts and Lyons [2020])

What was most intriguing about how visitors used APT in these different blends was how easily they switched back and forth from one blend to another and from first to third person and back again. This flexible use of APT was attributed by Ochs et al. [1996] as an element of expert practice, but we found it in even non-experts in this informal space as they engaged in multiple kinds of sensemaking moves. Though we do not attempt to imply causation from this correlation between APT and learning talk, we do think there is strong potential for further exploration into how embodiment and perspective taking can be harnessed to support learning interactions. For now, though, we turn to correlation and causation as their own area of exploration in embodied HDI.

8.2   Correlation and Causation

Our second case study is related to the concepts of correlation and causation—essential to understanding modern science (Bunge [1997]). After reviewing historic definitions of causation and correlation, we discuss two common challenges people have when looking for causation and correlation in data. We then illustrate how different prototypes of IDEA (Interactive Data and Embodied Analysis) altered peoples' sensemaking about correlation and causation in data (Alhakamy et al. [2021]). As we mention in Chapter 6, IDEA is an HDI installation for data exploration that we tested at Discovery Place Science, a science museum in Charlotte, NC—in a nutshell, two 3D globes representing geo-referenced data on a 65 in screen (Figure 6.5).

8.2.1   Historic Perspective on Causation and Correlation

In 1912, British philosopher Bertrand Russell delivered a paper ("On the Notion of Cause") at the Aristotelian Society in which he described causality as an erroneous, outdated concept that should have no place in science. According to Russell, the rules of "association" (i.e., "correlation") are all the laws that science can find, and causality cannot be inferred by symmetric association (Cartwright [1979]). Karl Pearson, one of the fathers of modern statistics, labeled causation in the 1911 edition of *Grammar of Science* as "another fetish amidst the inscrutable arcana of even modern science." This disillusion towards causation was grounded on the then new definition of "correlation." Francis Galton is considered the inventor of correlation in modern statistics: a pioneer in eugenics, Galton observed a "co-relation" between forearm length and height, head width and head breadth, and head length and height (Gillham [2001]). Galton defined "co-relation" as a phenomenon when "the variation of the one is accompanied on the average by more or less variation of the other, and in the same direction" (Galton [1889]). Karl Pearson further developed "correlation" into the mathematical model that we still use today.

Nowadays, however, there is a general consensus that casual relationships are an essential part of modern science (e.g., Suppes [1973]): controlled experiments are designed to identify causes and effects (Shadish et al. [2002]) by selecting dependent and independent variables and by limiting external influence thanks to randomization and the use of a control group (Dehue [2000]). Furthermore, the overwhelming evidence that correlation does not imply causation (e.g., Wolcott and Wolcott [1999]) contributed to the popular mantra that "correlation is not causation." This may be misleading, however, because it creates the assumption that correlation is not part of the scientific process; rather, correlation can be an important hint to causation, especially in well-designed controlled experiments that isolate the variables of interest (Novella [2009]).

8.2.2   Common Challenges while Interpreting Causation and Correlation

As a matter of fact, correlation and causation can be difficult to interpret. In particular, there are two common misconceptions.

- On the one hand, confounding correlation with causation may lead people to assume causality when there is not. For example, people may assume that high ice-cream sales directly increase the number of deaths by drowning (Moore [1996]), because of the strong correlation between these two variables. Rather, the reason for this correlation is a third variable (time), as they both increase during summer months (R. Johnson and Christensen [2019]).
- On the other hand, the mantra that correlation does not imply causation may conceal how correlation can be a precious hint to causation. This may lead people to question the validity of scientific findings, for example, the correlation between smoking and lung cancer despite the evidence on the strong association between the two (see Cornfield et al. [1959]).

This is where the design of the interaction can make a difference. We know from the theory of embodied schemata that people reason about correlation and causation using a variety of mental patterns (see Lakoff et al. [1999]). Full-body interaction may prime users towards a subset of these mental patterns (specifically, it may activate the "Causation is Movement of Location" metaphor), while different interaction styles may activate other schemata. In other words, the way in which we ask people to interact with the data visualization can have a priming effect similar to the one that is achieved with a priming frame in Framed Guessability. Thus, we may be able to direct the design in a way that limits the mistakes that people make when making sense of correlation and causation in data. This can complements existing work in data visualization that shows that visual elements and the type of data encoding (e.g., bar charts) affect people's perception of causation in data (Xiong et al. [2020]).

8.2.3   Comparing Two Interaction Styles: Full-Body vs. Gamepad

As we describe in detail in Alhakamy et al. [2021], we conducted an in-lab study using two alternative prototypes of IDEA in order to get a preliminary sense on how the design of HDI can aid people's sense making about correlation and causation. Participants in the first condition (Full-Body) were able to control the data visualization using mid-air gestures and body movements. Participants in the second condition (Gamepad) used a gamepad to interact with the screen. Twenty people participated in the study, 10 in each experimental conditions (see Figure 8.5).

Participants were asked to fill out a survey (pre-test) before interacting with IDEA. In the survey, they were asked to rate, on a scale from 1 to 5, how much they agreed with two statements that portrayed correlation and causation across the dataset on display (e.g., Fertilizer Consumption and Number of Threatened Fish Species). They were then asked to freely interact with the data

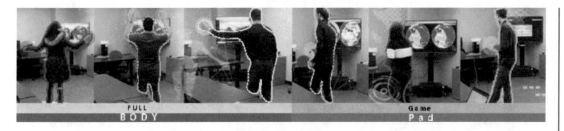

FIGURE 8.5: Participants in an in-lab study interact with two 3D globes representing geo-referenced data on a 65 in screen using either a gamepad or gestures and body movements.

visualization. After interacting with the HDI prototype, participants took the same survey (post-test).

The analysis of pre- and post-test scores using a two-way mixed ANOVA revealed a statistically significant two-way interaction between the interaction style (Full-Body vs. Gamepad) and the test time (the scores during pre-test vs. post-test), $F(1, 16) = 7.743, p = 0.013$. In other words, asking participants to interact using either gestures and body movements or a gamepad changed the average agreement scores that users assigned to causation and correlation statements about the data that they explored (Alhakamy et al. [2021]; see Figure 8.6).

An analysis of the remarks about the data on display that participants made while interacting with IDEA further revealed that people made more remarks based on FORCE schemata[5] in the Full-Body condition than in the Gamepad condition (Alhakamy et al. [2021]). This result is consistent with the way in which FORCE schemata are conceptualized according to Conceptual Metaphor Theory (CMT). In particular, FORCE schemata always involve a "sequence of causality" (Forceville [2016]): M. Johnson [2013] observes, for example, that "the door closes because I, or the wind, or a spring mechanism, acted on it to cause it to shut" (p. 44). The networks of metal connections (Lakoff and Johnson [2008]) between FORCE schemata and causality may explain the differences in the scores between experimental conditions. Further studies should investigate this phenomenon.

8.3 Analyzing the Interaction with HDI Installations

After reading about CoCensus and IDEA, a question may arise: How can we collect and analyze data to explore or evaluate people's interaction with HDI installations? We want to close this chapter with two methodological approaches that can structure the work of researchers and practitioners when conducting user studies with HDI installations. The first is Interaction Analysis (Jordan and

---

5. The reader may want to refer back to Section 3.2.

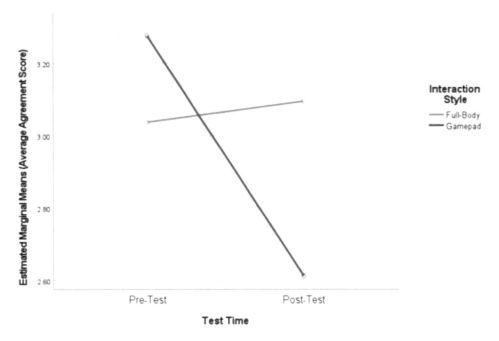

FIGURE 8.6: Plot of the estimated marginal means for the average agreement scores that participants assigned during pre-test vs. post-test in the two interaction styles (Full-Body vs. Gamepad).

Henderson [1995]), a method widely used in the learning sciences (Hall and Stevens [2015]) to analyze video recordings of people's interactions. The second is based on work by Hurtienne et al. [2015] on the use of Conceptual Metaphor Theory (CMT) for designing novel interactive systems.

8.3.1   Interaction Analysis

Interaction Analysis (IA) was defined in a seminal paper by Jordan and Henderson [1995] as "an interdisciplinary method for the empirical investigation of the interaction of human beings with each other and with objects in their environment." It includes structured indications on the data that should be collected, as well as on how to analyze and, if needed, transcribe video data.

8.3.1.1   Data Collection

The data collection process is based on ethnographic work (which includes observations, contextual inquiry, in-situ interviews)—to get an overall sense of the scenario and to identify the spots and activities to record—and on video recordings—which are then used to analyze gestures and interactions. The assumption here is that we cannot solely rely on ethnographic observations when studying gestures and body movements that people do: researchers may accidentally perceive subtle gestures (for example, somebody touching another person's shoulder when getting very close) that never actually occurred (Jordan and Henderson [1995]). Thus the need for video recordings.

8.3.1.2   Video Analysis

Multidisciplinary groups of researchers and practitioners view the tapes together and annotate it following a structured approach. Typically, the owners of the tape start playing it, and the tape "is stopped whenever a participant finds something worthy of remark" (Jordan and Henderson [1995], p. 44). At that point, the group makes observations (grounded on what was observed from the video) or hypotheses that can then be verified by observing similar patterns in other portions of the video. To avoid speculations that are no longer grounded on the videos, the video cannot be stopped for longer than five minutes. Hypotheses may include conjectures about mental states and mental events, for example, how well a person is able to understand an abstract concept that was presented to her given what she says, and her gestures and body movements.

When analyzing people's interaction with HDI installations, the mental states might be embodied schemata or embodied metaphors from Conceptual Metaphor Theory (CMT). For example, one could be interested in seeing if people frequently move their hand up and down when describing an increase or decrease in the density of a data variable.

On a side note, we want to highlight an interesting connection with Computer supported Cooperative Work (CSCW) literature, as pointed out in Hall and Stevens [2015]. Specifically, the notion of "boundary objects" (Star and Griesemer [1989]) is largely used in CSCW to denote (and study) objects that have a standardized meaning and that can be used to "coordinate the perspectives of various communities of practice" (Lee [2007]). Because videos are used in Interaction Analysis to support the work of heterogeneous, multi-disciplinary teams, they work as "boundary objects" across practitioners and researchers from different communities of practices (Hall and Stevens [2015]).

8.3.1.3   Transcription

The videos used in IA are not fully transcribed. If a portion of the video emerges as particularly relevant for the researchers' interest, it is typically transcribed. IA transcripts are not limited to what people say but also include annotations that emerged from the video analysis, along with other contextual information, such as body position, gaze, and gestures (Jordan and Henderson [1995]). The transcript is then analyzed by researchers, using methods that depend on their specific research questions.

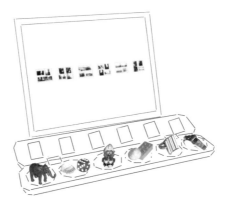

FIGURE 8.7: Tangible memory box designed in Hurtienne and Israel [2007]: tangible objects are used to represent and control collections of digital items on a screen.

8.3.1.4 Video Review Sessions

Sometimes, the study participants who were recorded are asked to review and discuss the videos with the researchers. This can be done to elicit additional information from participants or to validate/correct observations that emerged during the video analysis session (Jordan and Henderson [1995]).

8.3.2 Analyzing People's Interaction Using Conceptual Metaphor Theory

We highlighted Hurtienne's work when discussing Conceptual Metaphor Theory in Chapter 3. In particular, Table 3.1 reports the list of embodied schemata that were used in Hurtienne and Israel [2007] to present the "fictitious tangible interface sketch" of a memory box that collects items linked to digital media (see Figure 8.7).

Building on this work, Hurtienne et al. [2015] described an approach to design user interfaces that are intuitive (i.e., easy to use based on people's previous knowledge and experience), innovative, and inclusive for a wide range of age groups. Although Hurtienne et al. focuses on the design of a touch screen home entertainment system, the reader should notice how these three factors very well describe HDI installations, which are innovative, should be intuitive (as we describe in Chapter 7), and need to appeal to museum visitors that belong to different age groups (see Chapters 4 and 6).

The analysis process described in Hurtienne et al. [2015] is structured in five phases.

- *Contextual interviews*, conducted in-situ. This is the recommended approach to collect data for the CMT analysis. These interviews are then fully transcribed.
- *Interpretation session*. Members of the research team are trained on CMT.

- *Image schemata tagging.* A group of researchers scans the transcripts looking for keywords that point to embodied schemata. For example, the word "up" would be marked as an instance of the UP/DOWN schema. This process is done partially as a group and partially individually.
- *Image-schematic metaphor.* The research team reviews the embodied schemata that were coded in the transcripts looking for explicit ways in which such schemata were used metaphorically to describe abstract concepts. An example among those identified in Hurtienne et al. [2015] is: "Loud is Up, Soft is Down."
- *Metaphor clustering.* Image-schematic metaphors are then paired with requirements and functionalities of the interactive system.

Additionally, Hurtienne et al. [2015] describes design approaches that can be useful as a follow-up from the CMT analysis—particularly, when translating embodied schemata and metaphorical mappings into semi-functional and functional prototypes. These include affinity diagrams, wall walks (used, for example, to match metaphors with users' requirements), and user interviews with paper mock-ups.

The work on IDEA that we describe in Section 8.2 (Alhakamy et al. [2021] in particular) provides an example of how a similar type of analysis—based on CMT—allows exploration and assessment of the role of the interaction with an HDI installation in the way in which people make sense of data visualizations. Alhakamy et al. observe that, in the context of HDI, the use of embodied schemata may not be as literal as how it is reported in Hurtienne et al. [2015]. Thus, it is imperative to develop rigorous processes to assess inter-coder reliability (for example, by having multiple coders independently review portions of the coded transcripts), as well as to discuss and resolve disagreement. Additionally, Alhakamy et al. suggest the use of loglinear models to quantitatively assess differences in schematic reasoning across multiple participants.

Interestingly for museums, recent work in Soni et al. [2021] shows that CMT analysis is promising for exploring which types of gestural and bodily interactions may facilitate coordination and meaning-making in groups.

These two methodological approaches, however, barely scratch the surface of available techniques that could be adapted for analyzing embodied HDI systems. Regardless of the methods used, thorough analysis of embodied human-data interactions must account for a variety of mediating factors in the personal, physical, and sociocultural context of the interaction.

8.4   Summary: HDI for Sensemaking

The two case studies (CoCensus and IDEA) that we described in this chapter illustrate the opportunities for embodied HDI to facilitate data sensemaking.

Research on how people perceive and interpret data visualizations is not new. Back in 1987, for example, Cleveland and McGill conducted a series of experiments to identify how visual elements such as textures, colors, areas, and slope of the lines influence people's understanding of the data. More recently, Szafir [2017] discusses how colors are used to encode information and can facilitate data interpretation.

Similarly, gestures and body movements have been used in visual analytics (Thomas and Cook [2006]). For example, Ball and North [2007] conducted an experiment comparing a mouse-based vs. a full-body approach to navigate a data map on a wall display. Participants were faster in accomplishing their data analysis tasks (e.g., zooming in or aggregating data) when using the full-body approach to navigation.

The current work in HDI builds on these findings and indicates that the design of the interaction can facilitate unstructured (not task-based) exploration of large sets of data in informal learning settings.

CoCensus highlights opportunities to leverage perspective taking in order to facilitate collaborative sensemaking. IDEA shows that the interaction style (Gamepad vs. Full-Body) influences the way in which users agree with statements that portray correlation and causation across the data on display. By acknowledging and leveraging the unique affordances of physical movements, future work drawing on aspects of embodied design has great potential for expanding our understanding of HDI. The two methodological approaches that we reviewed at the end of this chapter (Interaction Analysis and designing with CMT) may structure this type of work.

CHAPTER 9

Conclusion

In this book, we explored the design of embodied human-data interactions, i.e., installations that facilitate exploration and sensemaking of large sets of data through embodied control. Because human-data interaction is a relatively recent sub-field in interaction design, however, there are still many exciting opportunities—and challenges—for research in the field. For this reason, we want to conclude with a look forward toward some of the open challenges that highlight promising research directions for HDI.

9.1 From Kinect to iPhone: Birth and Evolution of Off-the-Shelf Tracking Systems

Microsoft Kinect—the tracking camera that we frequently mention in this book—was launched in November 2010 as a gaming controller for the Xbox 360 console. It was initially a commercial success, with over 10 million units sold in the first four months. In late 2017, however, Microsoft officially discontinued manufacturing this camera. Most likely, a major issue was the lack of a "killer application": besides a few family-friendly titles, no major games popularized this tracking device.

Despite its lack of luck in the gaming industry, the Kinect has been largely used in research and in non-gaming applications, first with some "hacks" based on the Primesense tracking software, then directly using the Microsoft SDKs. Kinect cameras have been particularly helpful in embodiment research applications because they offer a relatively good accuracy of the tracking joints (about 10 mm; see Q. Wang et al. [2015]), combined with an unobtrusive setup (users are not required to wear markers to be tracked) and a relatively low price point (compared to more professionally oriented, custom-made tracking solutions).

In 2020, Microsoft released a new version of the Kinect, designed to be used with the Azure cloud platform rather than with a gaming console. Additionally, new small commercial tracking cameras are now available off-the-shelf: for example, Intel RealSense and Orbecc Astra.

Finally, the 2020 generation of iPhone Pros and IPads incorporates three cameras, one of which is an infra-red sensor based on Primesense technologies (Apple acquired Primesense in 2017). This basically turns these new phones and tablets into an ubiquitous version of the Kinect and opens up opportunities for further application scenarios using a device that many people already carry with them.

9.2   Future Research Directions for HDI

We hope in the past few chapters we have kindled your excitement for the potential of embodied HDI to foster engaging, productive, and meaningful learning experiences. We would be remiss, however, not to leave you with an outline of some of the challenges we, and others, have identified while designing and implementing HDI.

Take, for example, the CoCensus exhibit described in Chapter 8. We described the promising outcomes related to perspective taking and learning talk, but you may recall that the analysis was in the context of a 2x2 study design varying the *means of control*—full-body interaction or a handheld tablet controller—and the *distribution of control* between multiple users or a single global input (see Figure 8.4). Visitor dialogue from pairs of visitors interacting with one of the four versions of the exhibit was coded using the SQuILD method, allowing calculation of content scores that permitted statistical comparisons of the learning talk across conditions (J. Roberts and Lyons [2017a, 2017b]). Our own prior work, examples from other successful full-body exhibits, and theories of embodied cognition suggested that both full-body interaction and distributed control would productively mediate dialogue, with the synergistic combination of both those features leading to the highest amount of productive talk in the full-body, multi-input condition. However, our analysis revealed surprising results: the handheld sessions outperformed the full-body sessions on most metrics, in particular on SQuILD content scores. Moreover, the handheld single-input (global controller) condition presented the highest content scores of all (J. Roberts and Lyons [2017b]).

What was the cause of these surprising results? We further examined visitors' behavior in the exhibit to deduce whether significant differences in the amount of data viewed, the number of control actions taken, and the amount of time users spent learning the novel system could be the culprits, but none of these factors could explain the discrepancies. We had, then, to take a hard look at the limitations of off-the-desktop interactions.

9.2.1   The Need for Familiar Gestures

Surely, much fundamental and applied research still needs to be done before HDI can become mainstream. As we mentioned in Chapter 7, despite their tremendous potential to bring in patrons and to facilitate embodied learning, HDI installations have no utility if visitors are not able to use them. The problem of designing full-body interactions that feel natural to their users is particularly critical in museums, because visitors cannot consult user manuals before interacting with an installation. People tend to spend a short amount of time with an artifact or installation (often less than two minutes; see Ma et al. [2019]; Sandifer [1997]), and quickly leave thinking that the system is broken if the exhibit does not respond to their control actions.

On the contrary, touchscreens—whether personal phones or tablets or public kiosks—*are* quite mainstream. As a result, even small children are aware of the suite of control actions available on these devices: pinching to zoom, swiping to change screens or scroll, and tapping to select (Vatavu et al. [2015]).

Unfortunately, there are not yet many (if any) standardized gestures and body movements for mid-air interaction. With the CoCensus study we attempted to isolate the size of the gesture (and therefore the amount of the body involved in the control action) by creating parallel control actions for the two conditions. For example, a user might infer that if a tap action generates a system response on a screen, then a pinch or swipe action might also be valid inputs, but discovering a full-body input gesture (e.g., a large swipe of the arm) does not logically suggest other full-body controls. Sometimes, people tend to use legacy gestures (Morris et al. [2014]) when interacting with HDI installations—even when they are actually disruptive for the interaction. One of the works on the IDEA prototype that we discussed in Chapter 6, for example, describes how many museum visitors tried to touch a 65 in screen to control the data visualization, even if a Kinect camera was clearly visible below the screen (Trajkova et al. [2020a]). While none of the full-body control actions in our CoCensus study were inherently counter-intuitive or idiosyncratic, and they were all gestures suggested by visitors in early guessability studies (Cafaro et al. [2014a]), they do not always feel like part of a unified suite of actions that are a part of normal life like those on a touchscreen (J. Roberts and Lyons [2017b]).

Future work in HDI is in some sense reliant on broader adoption of embodied interaction across mainstream applications to allow users to build familiarity with suites of mid-air gestures. Alternatively, a promising approach—as proposed by Hurtienne [2017]—is to ground the design of gestures and body movements on shared, basic mental patterns (embodied schemata) that make such controlling actions look familiar (even if they are novel).

9.2.2 Embodiment without Touch?

It must be noted that two of the three control actions in the full-body conditions of CoCensus were touchless. Swipes were performed in the air, and while some visitors finished the aggregation gesture of drawing their arms together in an exaggerated pinch with a clap, this clap was not a necessary part of the control action.

The tendency of visitors to do this clap may speak to the desire for some sort of haptic feedback for these control actions. The field of tangible, embedded, and embodied interactions is based on the idea of building systems that allow humans and computers to interact with fewer intermediaries, using "increased physical engagement and direct interaction" (Hornecker [2011]). While the full-body design of CoCensus and IDEA did have fewer intermediaries (i.e., the removal of the external controller), it could be argued that the increased size of the control actions did not actually lead to "increased physical engagement." After all, what is more physical than touch?

Chattopadhyay and Bolchini [2013] examine touchless interfaces in wall-sized displays and emphasize the importance of visual feedback in absence of haptic feedback. The observed differences between full-body and handheld conditions in learning talk suggest that the full-body version of CoCensus as tested here may not have adequately accommodated for this lack of touch.

9.2.3   Defining Schemata

In Chapter 3 we provided an overview of Conceptual Metaphor Theory (CMT), and in Chapter 8 we discussed an example of how CMT can be used during thematic analysis and coding to interpret the effect of the interaction on the way in which people see correlation and causation in data. The process of identifying embodied schemata (and metaphors) in interview or think-aloud transcripts is, however, challenging. For example, while labels like CONTAINER, COUNTERFORCE, DIVERSION (see Chapter 3) may seem intuitive at a first glance, using them to code people's conversation around HDI installations is a non-trivial task. In other words, there is not yet a common dictionary of keywords that allows researchers to identify specific schemata, because much depends on the linguistic context. The creation of such a standardized dictionary may facilitate the task of effectively using CMT for the design of HDI going forward.

9.2.4   Assessing Learning

In Chapter 8, we discussed how the design of the interaction may alter data sensemaking, as well as the conversations that people have around interactive data visualizations. The work that we discussed in this book, however, has just started to tackle the interplay between embodied interaction, data visualization, and learning. For example, research in HDI still needs to identify common approaches to evaluate the effect of the interaction on medium- and long-term learning. Our hope is that this book will be able to chart the multi-disciplinary foundations of HDI to a community of fellow researchers and start a conversation about new application scenarios and opportunities for convergent research.

9.3   HDI in the Wild

As we described in Chapter 7, most elicitation-based design methodologies focus on crafting controlling actions in-lab rather than collecting data in the intended context of use (as suggested, for example, at the end of Trajkova et al. [2020a]). This is a sharp contrast to the idea of embodied interaction (Dourish [2001]) that we discussed in Chapter 3: the actions that people do (or recommend) in front of an interactive installation may have different meaning depending on the surrounding space and the social context. As we increasingly focus on developing data interactions for organic, social, informal, and free-choice environments, continued attention to the composition and motivation of user groups, the perspectives they employ, the display's content, and the design of both the physical and digital spaces will help us explore not only what embodied HDI is good for but where it falls short.

Conversations on the designs of educational technologies tend to fall toward one of two perspectives. A *techno-centric* perspective looks at new technologies as the solutions to problems. As Fishman et al. [2016] explain, "A techno-centric perspective, for example, would ask whether the introduction of a technology, such as digital whiteboards or "clicker" response systems, helps students learn more effectively." We (along with Fishman et al.) encourage HDI designers and researchers to adopt a *socio-technical* approach to their work "that views the products of technology use as emerging from interactions among social and organizational structures, people, and tools" (Fishman et al. [2016]). That is, data interactions are no longer situated in isolated, single-user desktops or lab settings but are part of a complex, sociocultural, "in the wild" system.

Humans rely on data to make decisions across virtually every aspect of our lives. We frequently employ graphs, maps, and other spatialized forms to aid data interpretation, yet the familiarity of these displays causes us to forget that even basic representations are complex, challenging inscriptions and are not neutral; they are based on representational choices that impact how and what they communicate. Embodiment is one tool to mediate how people learn through, with, and about visual representations of data. The multi-disciplinary, situated perspective on embodied HDI that we present in this book demonstrates how advances in off-the-shelf sensing technologies have the potential to support data sensemaking, increase data interactions by non-scientific public audiences, and take the next (literal) step towards the democratization of data.

Bibliography

Ackad, Christopher, Martin Tomitsch, and Judy Kay. Skeletons and silhouettes: Comparing user representations at a gesture-based large display. In *Proceedings of the 2016 CHI Conference on Human Factors in Computing Systems (CHI'16)*, pages 2343–2347, New York, NY, USA, 2016. Association for Computing Machinery. ISBN 9781450333627. DOI: 10.1145/2858036.2858427.

Alhakamy, A'aeshah, Milka Trajkova, and Francesco Cafaro. Show me how you interact, I will tell you what you think: Exploring the effect of the interaction style on users' sensemaking about correlation and causation in data. In *Designing Interactive Systems Conference 2021 (DIS'21)*, pages 564–575. New York, NY, USA, 2021. Association for Computing Machinery. DOI: 10.1145/3461778.3462083.

Alhakamy, A'aeshah, F. Cafaro, M. Trajkova, S. Kankara, R. Mallappa, and S. Veda. Design strategies and optimizations for human-data interaction systems in museums. In *2020 IEEE 20th International Conference on Advanced Learning Technologies (ICALT)*, pages 246–248, 2020.

Ali, Abdullah, Meredith Ringel Morris, and Jacob O. Wobbrock. "i am iron man": Priming improves the learnability and memorability of user-elicited gestures. In *Proceedings of the 2021 CHI Conference on Human Factors in Computing Systems (CHI'21)*, New York, NY, USA, 2021. Association for Computing Machinery. ISBN 9781450380966. DOI: 10.1145/3411764.3445758.

Ali, Abdullah, Meredith Ringel Morris, and Jacob O. Wobbrock. Crowdlicit: A system for conducting distributed end-user elicitation and identification studies. In *Proceedings of the 2019 CHI Conference on Human Factors in Computing Systems (CHI'19)*, pages 1–12, New York, NY, USA, 2019. Association for Computing Machinery. ISBN 9781450359702. DOI: 10.1145/3290605.3300485.

Allen, Sue. Looking for learning in visitor talk: A methodological exploration. In *Learning conversations in museums*, pages 265–309. Routledge, 2003.

Allen, Sue, and Joshua Gutwill. Designing with multiple interactives: Five common pitfalls. *Curator: The Museum Journal*, 47(2):199–212, 2004.

Allen, Sue, and Joshua Gutwill. Creating a program to deepen family inquiry at interactive science exhibits. *Curator: The Museum Journal*, 52(3):289–306, 2009.

Andrews, Christopher, Alex Endert, and Chris North. *Space to think: Large high-resolution displays for sensemaking*, In *Proceedings of the SIGCHI Conference on Human Factors in Computing Systems (CHI'10)*, pages 55–64, New York, NY, USA, 2010. Association for Computing Machinery. DOI: 10.1145/1753326.1753336.

Antle, Alissa N., Greg Corness, Saskia Bakker, Milena Droumeva, Elise van den Hoven, and Allen Bevans. Designing to support reasoned imagination through embodied metaphor. In *Proceedings of the Seventh ACM Conference on Creativity and Cognition (C&C'09)*, pages 275–284, New York, NY, USA, 2009. Association for Computing Machinery. DOI: 10.1145/1640233.1640275.

Ash, Doris. Dialogic inquiry in life science conversations of family groups in a museum. *Journal of Research in Science Teaching*, 40(2):138–162, 2003.

Atkins, Leslie J., Lisanne Velez, David Goudy, and Kevin N Dunbar. The unintended effects of interactive objects and labels in the science museum. *Science Education*, 93(1):161–184, 2009.

Ball, Robert, and Chris North. Realizing embodied interaction for visual analytics through large displays. *Computers & Graphics*, 31(3):380–400, 2007. ISSN 0097-8493. DOI: 10.1016/j.cag.2007.01.029.

Bao, Patti, and Darren Gergle. What's "this" you say? The use of local references on distant displays. In *Proceedings of the SIGCHI Conference on Human Factors in Computing Systems (CHI'09)*, pages 1029–1032, New York, NY, USA, 2009. Association for Computing Machinery. DOI: 10.1145/1518701.1518858.

Beheshti, Elham, Anne Van Devender, and Michael Horn. Touch, click, navigate: Comparing tabletop and desktop interaction for map navigation tasks. In *Proceedings of the 2012 ACM international conference on Interactive tabletops and surfaces (ITS'12)*, pages 205–214, New York, NY, USA, 2012. Association for Computing Machinery. DOI: 10.1145/2396636.2396669.

Bell, Benjamin, Ray Bareiss, and Richard Beckwith. Sickle cell counselor: A prototype goal-based scenario for instruction in a museum environment. *Journal of the Learning Sciences*, 3(4):347–386, 1994.

Bellotti, Francesco, C. Berta, Alessandro De Gloria, and Massimiliano Margarone. User testing a hypermedia tour guide. *IEEE Pervasive Computing*, 1(2):33–41, 2002.

Berman, Francine, Rob Rutenbar, Brent Hailpern, Henrik Christensen, Susan Davidson, Deborah Estrin, Michael Franklin, Margaret Martonosi, Padma Raghavan, Victoria Stodden, and Alexander S. Szalay Realizing the potential of data science. *Communications of the ACM*, 61 (4):67–72, 2018.

Bernard, Russell H. Research methods in anthropology-qualitative and quantitative approaches. Technical Report. Altamira Press, 2002.

Birchfield, David, Thomas Ciufo, and Gary Minyard. SMALLab: A mediated platform for education. In *ACM SIGGRAPH 2006 Educators Program (SIGGRAPH'06)*, pages 33–es, New York, NY, USA, 2006. Association for Computing Machinery. DOI: 10.1145/1179295.1179329.

Birchfield, David, Ellen Campana, Sarah Hatton, Mina Johnson-Glenberg, Aisling Kelliher, Loren Olson, Christopher Martinez, Philippos Savvides, Lisa Tolentino, and Sibel Uysal. Embodied and mediated learning in SMALLab: A student-centered mixed-reality environment. In *ACM SIGGRAPH 2009 Emerging Technologies (SIGGRAPH'09)*, article 9, 1, New York, NY, USA, 2009. ACM. ISBN 9781605588339. DOI: 10.1145/1597956.1597965.

Bitgood, Stephen, and Harris H. Shettel. An overview of visitor studies. *Journal of Museum Education*, 21(3):6–10, 1996.

Block, Florian, Michael S. Horn, Brenda Caldwell Phillips, Judy Diamond, E. Margaret Evans, and Chia Shen. The deeptree exhibit: Visualizing the tree of life to facilitate informal learning. *IEEE Transactions on Visualization and Computer Graphics*, 18(12):2789–2798, 2012. DOI: 10.1109/TVCG.2012.272.

Börner, Katy, Adam Maltese, Russell Nelson Balliet, and Joe Heimlich. Investigating aspects of data visualization literacy using 20 information visualizations and 273 science museum visitors. *Information Visualization*, 15(3):198–213, 2016.

Borun, Minda, Margaret Chambers, and Ann Cleghorn. Families are learning in science museums. *Curator: The Museum Journal*, 39(2):123–138, 1996.

Bransford, John D., Ann L. Brown, Rodney R. Cocking. *How people learn*, volume 11. National Academy Press, 2000.

Brignull, Harry, and Yvonne Rogers. Enticing people to interact with large public displays in public spaces. In *Proceedings of INTERACT (INTERACT'03)*, 3:17–24 ISO Press, 2003.

Brunyé, Tad T., Tali Ditman, Caroline R. Mahoney, Jason S. Augustyn, and Holly A. Taylor. When you and I share perspectives: Pronouns modulate perspective taking during narrative comprehension. *Psychological Science*, 20(1):27–32, 2009.

Bunge, Mario. Mechanism and explanation. *Philosophy of the Social Sciences*, 27(4):410–465, 1997.

Cabitza, Federico, Angela Locoro, Daniela Fogli, and Massimiliano Giacomin. Valuable visualization of healthcare information: From the quantified self data to conversations. In *Proceedings of the International Working Conference on Advanced Visual Interfaces (AVI 16)*, pages 376–380, New York, 2016. ACM. ISBN 9781450341318. DOI: 10.1145/2909132.2927474.

Cafaro, Francesco. Using embodied allegories to design gesture suites for human-data interaction. In *Proceedings of the 2012 ACM Conference on Ubiquitous Computing (UbiComp'12)*, page 560, New York, NY, USA, 2012. ACM Press. ISBN 9781450312240. DOI: 10.1145/2370216.2370309.

Cafaro, Francesco. *Using framed guessability to design gesture suites for embodied interaction.* Ph.D. thesis, University of Illinois Urbana-Champaign, 2015.

Cafaro, Francesco, Leilah Lyons, and Alissa N. Antle. Framed guessability: Improving the discoverability of gestures and body movements for full-body interaction. In *Proceedings of the 2018 CHI Conference on Human Factors in Computing Systems (CHI'18)*, pages 1–12, 2018.

Cafaro, Francesco, Leilah Lyons, Raymond Kang, Josh Radinsky, Jessica Roberts, and Kristen Vogt. Framed guessability: Using embodied allegories to increase user agreement on gesture sets. In *Proceedings of the 8th International Conference on Tangible, Embedded and Embodied Interaction (TEI'14)*, pages 197–204, New York, NY, USA, 2014a. Association for Computing Machinery. ISBN 9781450326353. DOI: 10.1145/2540930.2540944.

Cafaro, Francesco, Leilah Lyons, Jessica Roberts, and Josh Radinsky. The uncanny valley of embodied interaction design. In *Proceedings of the 2014 Conference on Designing Interactive Systems (DIS'14)*, pages 1075–1078, New York, NY, USA, 2014b. Association for Computing Machinery. DOI: 10.1145/2598510.2598593.

Cafaro, Francesco, Alessandro Panella, Leilah Lyons, Jessica Roberts, and Josh Radinsky. I see you there! Developing identity-preserving embodied interaction for museum exhibits. In *Proceedings of the SIGCHI Conference on Human Factors in Computing Systems (CHI'13)*, pages 1911–1920, New York, NY, USA, 2013. Association for Computing Machinery. ISBN 9781450318990. DOI: 10.1145/2470654.2466252.

Card, Stuart K., Jock D. Mackinlay, and Ben Shneiderman *Readings in information visualization: Using vision to think.* Morgan Kaufmann, 1999.

Cartwright, Nancy. Causal laws and effective strategies. In *Noûs*, pages 419–437. Wiley, 1979.

Castellucci, Steven J., and I. Scott MacKenzie. Graffiti vs. unistrokes: An empirical comparison. In *Proceedings of the SIGCHI Conference on Human Factors in Computing Systems (CHI'08)*, pages 305–308, New York, NY, USA, 2008. Association for Computing Machinery. ISBN 9781605580111. DOI: 10.1145/1357054.1357106.

Charmaz, Kathy, and Linda Liska Belgrave. Grounded theory. *The Blackwell Encyclopedia of Sociology*, 2007. DOI: 10.1002/9781405165518.wbeosg070.pub2.

Chattopadhyay, Debaleena, and Davide Bolchini. Laid-back, touchless collaboration around wall-size displays: Visual feedback and affordances. 2013. Retrieved online at: https://scholar works.iupui.edu/handle/1805/4526.

Cheung, Victor, Diane Watson, Jo Vermeulen, Mark Hancock, and Stacey Scott. Overcoming interaction barriers in large public displays using personal devices. In *Proceedings of the Ninth ACM International Conference on Interactive Tabletops and Surfaces (ITS'14)*, pages 375–380, New York, NY, USA, 2014. Association for Computing Machinery. ISBN 9781450325875. DOI: 10.1145/2669485.2669549.

Chi, Michelene T. H. Quantifying qualitative analyses of verbal data: A practical guide. *Journal of the Learning Sciences*, 6(3):271–315, 1997.

Clarke, Loraine, Eva Hornecker, and Ian Ruthven. Fighting fires and powering steam locomotives: Distribution of control and its role in social interaction at tangible interactive museum exhibits. In *Proceedings of the 2021 CHI Conference on Human Factors in Computing Systems (CHI'21)*, New York, NY, USA, 2021. Association for Computing Machinery. ISBN 9781450380966. DOI: 10.1145/3411764.3445534.

Cleveland, William S., and Robert McGill. Graphical perception: The visual decoding of quantitative information on graphical displays of data. *Journal of the Royal Statistical Society: Series A (General)*, 150(3):192–210, 1987.

Coenen, Jorgos, Sandy Claes, and Andrew Vande Moere. The concurrent use of touch and mid-air gestures or floor mat interaction on a public display. In *Proceedings of the 6th ACM International Symposium on Pervasive Displays*, page 9, USA, 2017. Association for Computing Machinery.

Corbin, Juliet, and Anselm Strauss. *Basics of qualitative research: Techniques and procedures for developing grounded theory*. Sage, 2014.

Cornfield, Jerome, William Haenszel, E. Cuyler Hammond, Abraham M. Lilienfeld, Michael B. Shimkin, and Ernst L. Wynder. Smoking and lung cancer: Recent evidence and a discussion of some questions. *Journal of the National Cancer Institute*, 22(1):173–203, 1959.

Correia, Nuno, Tarquínio Mota, Rui Nóbrega, Luís Silva, and Andreia Almeida. A multi-touch tabletop for robust multimedia interaction in museums. In *ACM International Conference on Interactive Tabletops and Surfaces (ITS'10)*, pages 117–120, New York, NY, USA, 2010. Association for Computing Machinery.

Crowder, Elaine M. Gestures at work in sense-making science talk. *Journal of the Learning Sciences*, 5(3):173–208, 1996.

Crowley, Kevin, and Melanie Jacobs. Building islands of expertise in everyday family activity. *Learning conversations in museums*, 333–356. Routledge, 2002.

Curcio, Frances R. Comprehension of mathematical relationships expressed in graphs. *Journal for Research in Mathematics Education*, 18(5): 382–393, 1987. retrieved online at: https://pubs.nctm.org/view/journals/jrme/18/5/article-p382.xml.

Danish, Joshua A., Noel Enyedy, Asmalina Saleh, Christine Lee, and Alejandro Andrade. Science through technology enhanced play: Designing to support reflection through play and embodiment. *Exploring the Material Conditions of Learning: The Computer Supported Collaborative Learning (CSCL) Conference 2015*. Gothenburg, Sweden: The International Society of the Learning Sciences, 2015.

Danish, Joshua A., Noel Enyedy, Asmalina Saleh, and Megan Humburg. Learning in embodied activity framework: A sociocultural framework for embodied cognition. *International Journal of Computer-Supported Collaborative Learning*, 15(1):49–87, 2020.

Davis, Pryce, Michael Horn, Laurel Schrementi, Florian Block, Brenda Phillips, E. Margaret Evans, Judy Diamond, and Chia Shen. Going deep: Supporting collaborative exploration of evolution in natural history museums. *To See the World and a Grain of Sand: Learning across Levels of Space, Time, and Scale: CSCL 2013 Conference Proceedings Volume 1—Full Papers & Symposia*, pages 153–160. Madison, WI: International Society of the Learning Sciences, 2013.

Dehue, Trudy. From deception trials to control reagents: The introduction of the control group about a century ago. *American Psychologist*, 55(2):264, 2000.

Diamond, Judy, Michael Horn, and David H. Uttal. *Practical evaluation guide: Tools for museums and other informal educational settings*. Rowman & Littlefield, 2016.

DiPaola, Steve, and Caitlin Akai. Designing an adaptive multimedia interactive to support shared learning experiences. In *ACM SIGGRAPH 2006 Educators Program (SIGGRAPH'06)*, pages 14–es. New York, NY, USA, 2006. Association for Computing Machinery. DOI: 10.1145/1179295.1179310.

Donoho, A. W., D. L. Donoho, and M. Gasko. Macspin: Dynamic graphics on a desktop computer. *IEEE Computer Graphics and Applications*, 8(4):51–58, 1988.

Dourish, Paul. *Where the action is*. MIT Press, 2001.

Dourish, Paul. Epilogue: Where the action was, wasn't, should have been, and might yet be. *ACM Transactions on Computer-Human Interaction (TOCHI)*, 20(1):1–4, 2013.

Dreyer, Felix R., Dietmar Frey, Sophie Arana, Sarah von Saldern, Thomas Picht, Peter Vajkoczy, and Friedemann Pulvermüller. Is the motor system necessary for processing action and abstract emotion words? Evidence from focal brain lesions. *Frontiers in Psychology*, 6:1661, 2015.

Dutsch, Dorota. Towards a grammar of gesture: A comparison between the types of hand movements of the orator and the actor in quintilian's institutio oratoria 11.3.85–184. *Gesture*, 2(2): 259–281, 2002.

Eberbach, Catherine, and Kevin Crowley. From living to virtual: Learning from museum objects. *Curator: The Museum Journal*, 48(3):317–338, 2005.

Edelson, Daniel C., Douglas N. Gordin, and Roy D. Pea. Addressing the challenges of inquiry-based learning through technology and curriculum design. *Journal of the Learning Sciences*, 8 (3–4):391–450, 1999.

Elmqvist, Niklas. Embodied human-data interaction. In *ACM CHI 2011 Workshop "Embodied Interaction: Theory and Practice in HCI,"* pages 104–107, USA, 2011. Association for Computing Machinery.

England, David. Whole body interaction: An introduction. In *Whole body interaction*, pages 1–5. Springer, 2011.

England, David, Martin Randles, Paul Fergus, and A. Taleb-Bendiab. Towards an advanced framework for whole body interaction. In *Virtual and mixed reality*, pages 32–40, Berlin, Heidelberg, ISBN 9783642027710.

Enyedy, Noel, Joshua Danish, and David DeLiema. Constructing and deconstructing materially-anchored conceptual blends in an augmented reality collaborative learning environment. In *To See the World and a Grain of Sand: Learning across Levels of Space, Time, and Scale: CSCL 2013 Conference Proceedings Volume 1—Full Papers & Symposia*, pages 192–199. Madison, WI: International Society of the Learning Sciences, 2013.

Erickson, Thomas, David N. Smith, Wendy A. Kellogg, Mark Laff, John T. Richards, and Erin Bradner. Socially translucent systems: Social proxies, persistent conversation, and the design of "babble." In *Proceedings of the SIGCHI conference on Human Factors in Computing Systems (CHI'99)*, pages 72–79, New York, NY, USA, 1999. Association for Computing Machinery. DOI: 10.1145/302979.302997.

Ertmer, Peggy A., and Timothy J. Newby. Behaviorism, cognitivism, constructivism: Comparing critical features from an instructional design perspective. *Performance Improvement Quarterly*, 26(2):43–71, 2013.

Falcão, Taciana Pontual, and Sara Price. What have you done! The role of "interference" in tangible environments for supporting collaborative learning. In *CSCL*, (1):325–334, 2009.

Falk, John H. An identity-centered approach to understanding museum learning. *Curator: The Museum Journal*, 49(2):151–166, 2006.

Falk, John H., and Lynn D. Dierking. *Learning from museums.* 2nd ed. Rowman & Littlefield, 2018.

Falk, John H., Theano Moussouri, and Douglas Coulson. The effect of visitors' agendas on museum learning. *Curator: The Museum Journal*, 41(2):107–120, 1998.

Few, Stephen. *Now you see it: Simple visualization techniques for quantitative analysis.* (1st. ed.). Oakland, CA: Analytics Press, 2009.

Filipi, Anna, and Roger Wales. Perspective-taking and perspective-shifting as socially situated and collaborative actions. *Journal of Pragmatics*, 36(10):1851–1884, 2004.

Fillmore, Charles J. Frame semantics. *Cognitive Linguistics: Basic Readings*, 34:373–400, 2006.

Fischer, Martin H., Nele Dewulf, and Robin L. Hill. Designing bar graphs: Orientation matters. *Applied Cognitive Psychology: The Official Journal of the Society for Applied Research in Memory and Cognition*, 19(7):953–962, 2005.

Fishman, Barry, Chris Dede, and Barbara Means. Teaching and technology: New tools for new times. *Handbook of research on teaching*, pages 1269–1334. American Educational Research Association, 2016. DOI: 10.3102/978-0-935302-48-6_21.

Forceville, Charles. The force and balance schemas in journey metaphor animations. *Multimodality and performance*, pages 8–22. Cambridge Scholars Publishing, 2016.

Forlines, Clifton, and Chia Shen. Dtlens: Multi-user tabletop spatial data exploration. In *Proceedings of the 18th Annual ACM Symposium on User Interface Software and Technology (UIST'05)*, pages 119–122, New York, NY, USA, 2005. Association for Computing Machinery. DOI: 10.1145/1095034.1095055.

Forlines, Clifton, Alan Esenther, Chia Shen, Daniel Wigdor, and Kathy Ryall. Multi-user, multi-display interaction with a single-user, single-display geospatial application. In *Proceedings of the 19th Annual ACM Symposium on User Interface Software and Technology (UIST'06)*, pages 273–276, New York, NY, USA, 2006. Association for Computing Machinery. DOI: 10.1145/1166253.1166296.

Frank, Mark, Johanna Walker, Judie Attard, and Alan Tygel. Data literacy—what is it and how can we make it happen? *Journal of Community Informatics*, 12(3), 2016.

Friel, Susan N., Frances R. Curcio, and George W. Bright. Making sense of graphs: Critical factors influencing comprehension and instructional implications. *Journal for Research in Mathematics Education*, 32(2):124–158, 2001.

Galton, Francis. I. Co-relations and their measurement, chiefly from anthropometric data. *Proceedings of the Royal Society of London*, 45(273–279):135–145, 1889.

Gentner, Dedre, and Donald R. Gentner. Flowing waters or teeming crowds: Mental models of electricity. In *Mental models*, pages 107–138. Psychology Press, 2014.

Gillham, Nicholas Wright. *A life of Sir Francis Galton: From African exploration to the birth of eugenics*. Oxford University Press, 2001.

Glazer, Nirit. Challenges with graph interpretation: A review of the literature. *Studies in Science Education*, 47(2):183–210, 2011.

Goldman, Susan R., Anthony J. Petrosino, et al. Design principles for instruction in content domains: Lessons from research on expertise and learning. *Handbook of applied cognition*, pages 595–628. Wiley & Sons, 1999.

Good, Michael D., John A. Whiteside, Dennis R. Wixon, and Sandra J. Jones. Building a user-derived interface. *Communications of the ACM*, 27(10):1032–1043, 1984.

Gotz, David, Shun Sun, and Nan Cao. Adaptive contextualization: Combating bias during high-dimensional visualization and data selection. In *Proceedings of the 21st International Conference on Intelligent User Interfaces (IUI'16)*, pages 85–95, New York, NY, USA, 2016. Association for Computing Machinery. ISBN 9781450341370. DOI: 10.1145/2856767.2856779.

Greenberg, Saul, Nicolai Marquardt, Till Ballendat, Rob Diaz-Marino, and Miaosen Wang. Proxemic interactions: The new ubicomp? *Interactions*, 18(1):42–50, January 2011. ISSN 1072-5520. DOI: 10.1145/1897239.1897250.

Greeno, James G., Allan M Collins, Lauren B Resnick. Cognition and learning. *Handbook of Educational Psychology*, 77:15–46, 1996.

Gutwill, Joshua P. Gaining visitor consent for research ii: Improving the posted-sign method. *Curator: The Museum Journal*, 46(2):228–235, 2003.

Gutwill, Joshua P., and Sue Allen. Deepening students' scientific inquiry skills during a science museum field trip. *Journal of the Learning Sciences*, 21(1):130–181, 2012.

Hall, Rogers, and Reed Stevens. Interaction analysis approaches to knowledge in use. In *Knowledge and interaction: A synthetic agenda for the learning sciences*, pages 72–108, 2015.

Harpaintner, Marcel, Eun-Jin Sim, Natalie M. Trumpp, Martin Ulrich, and Markus Kiefer. The grounding of abstract concepts in the motor and visual system: An fmri study. *Cortex*, 124: 1–22, 2020. DOI: 10.1016/j.cortex.2019.10.014.

Heath, Christian, and Vom Lehn Dirk. Configuring "interactivity" enhancing engagement in science centres and museums. *Social Studies of Science*, 38(1):63–91, 2008.

Heer, Jeffrey, Frank Van Ham, Sheelagh Carpendale, Chris Weaver, and Petra Isenberg. Creation and collaboration: Engaging new audiences for information visualization. In *Information visualization*, pages 92–133. Springer, 2008.

Hein, George E. *Learning in the museum*. Routledge, 2002.

Hill, Michael J., and Richard J Aspinall. *Spatial information for land use management*. CRC Press, 2000.

Hinrichs, U., H. Schmidt, and S. Carpendale. Emdialog: Bringing information visualization into the museum. *IEEE Transactions on Visualization and Computer Graphics*, 14(6):1181–1188, 2008. DOI: 10.1109/TVCG.2008.127.

Horn, Michael, Zeina Atrash Leong, Florian Block, Judy Diamond, E. Margaret Evans, Brenda Phillips, and Chia Shen. Of bats and apes: An interactive tabletop game for natural history museums. In *Proceedings of the SIGCHI Conference on Human Factors in Computing Systems*, pages 2059–2068, New York, NY, USA, 2012. Association for Computing Machinery.

Hornecker, Eva. "I don't understand it either, but it is cool"—Visitor interactions with a multi-touch table in a museum. In *2008 3rd IEEE International Workshop on Horizontal Interactive Human Computer Systems*, pages 113–120, 2008.

Hornecker, Eva. Interactions around a contextually embedded system. In *Proceedings of the Fourth International Conference on Tangible, Embedded, and Embodied Interaction (TEI'11)*, pages 169–176, New York, NY, USA, 2010. Association for Computing Machinery.

Hornecker, Eva. The role of physicality in tangible and embodied interactions. *Interactions*, 18(2): 19–23, 2011.

Hornecker, Eva, and Jacob Buur. Getting a grip on tangible interaction: A framework on physical space and social interaction. In *Proceedings of the SIGCHI Conference on Human Factors in Computing Systems*, pages 437–446, New York, NY, USA, 2006. Association for Computing Machinery.

Hornecker, Eva, and Luigina Ciolfi. Human-computer interactions in museums. *Synthesis Lectures on Human-Centered Informatics*, 12(2):i–171, 2019.

Hsi, Sherry. A study of user experiences mediated by nomadic web content in a museum. *Journal of Computer Assisted Learning*, 19(3):308–319, 2003.

Huang, Elaine M., Anna Koster, and Jan Borchers. Overcoming assumptions and uncovering practices: When does the public really look at public displays? In *International Conference on Pervasive Computing*, pages 228–243. Springer, 2008.

Hurtienne, Jörn. How cognitive linguistics inspires hci: Image schemas and image-schematic metaphors. *International Journal of Human-Computer Interaction*, 33(1):1–20, 2017.

Hurtienne, Jörn, and Johann Habakuk Israel. Image schemas and their metaphorical extensions: Intuitive patterns for tangible interaction. In *Proceedings of the 1st International Conference on Tangible and Embedded Interaction (TEI'07)*, pages 127–134, New York, NY, USA, 2007. Association for Computing Machinery.

Hurtienne, Jörn, Kerstin Klöckner, Sarah Diefenbach, Claudia Nass, and Andreas Maier. Designing with image schemas: Resolving the tension between innovation, inclusion and intuitive use. *Interacting with Computers*, 27(3):235–255, 2015.

Hurtienne, Jörn, Franzisca Maas, Astrid Carolus, Daniel Reinhardt, Cordula Baur and Carolin Wienrich. Move&Find: The value of kinaesthetic experience in a casual data representation. IEEE Computer Graphics and Applications, 40(6):61–75, 2020. DOI: 10.1109/MCG.2020.3025385.

Husserl, Edmund. *Logical investigations*, volume 1. Psychology Press, 2001.

Hutchins, Edwin L., James D. Hollan, and Donald A. Norman. Direct manipulation interfaces. *Human–Computer Interaction*, 1(4):311–338, 1985.

Ihde, Don. Technology and the lifeworld: From garden to earth. Indiana University Press, 1990.

Ishii, Hiroshi. The tangible user interface and its evolution. *Communications of the ACM*, 51(6): 32–36, 2008.

Ishii, Hiroshi, and Brygg Ullmer. Tangible bits: Towards seamless interfaces between people, bits and atoms. In *Proceedings of the ACM SIGCHI Conference on Human Factors in Computing*

Systems (CHI'97), pages 234–241, New York, NY, USA, 1997. Association for Computing Machinery. DOI: 10.1145/258549.258715.

Jackson, Bret, Tung Yuen Lau., David Schroeder, Kimani C. Toussaint, and Daniel F. Keefe. A lightweight tangible 3D interface forinteractive visualization of thin fiber structures. IEEE Transactions on Visualization and Computer Graphics, 19(12):2802–2809, 2013. DOI: 10.1109/TVCG.2013.121.

Jacobs, Jennifer K., Makoto Yoshida, James W. Stigler, and Clea Fernandez. Japanese and american teachers' evaluations of mathematics lessons: A new technique for exploring beliefs. *Journal of Mathematical Behavior*, 16(1):7–24, 1997.

Johnson, Mark. *The body in the mind: The bodily basis of meaning, imagination, and reason*. University of Chicago Press, 2013.

Johnson, R. Burke, and Larry Christensen. *Educational research: Quantitative, qualitative, and mixed approaches*. Sage, 2019.

Johnson-Glenberg, Mina C., David Birchfield, Lisa Tolentino, and Tatyana Koziupa. Collaborative embodied learning in mixed reality motion-capture environments: Two science studies. *Journal of Educational Psychology*, 106(1):86, 2014.

Johnson-Glenberg, Mina C., David Birchfield, and Sibel Usyal. Smallab: Virtual geology studies using embodied learning with motion, sound, and graphics. *Educational Media International*, 46(4):267–280, 2009.

Johnston, Hank. Verification and proof in frame and discourse analysis. *Methods of Social Movement Research*, 16:62–91, 2002.

Jordan, Brigitte, and Austin Henderson. Interaction analysis: Foundations and practice. *Journal of the Learning Sciences*, 4(1):39–103, 1995.

Kaptelinin, Victor. The mediational perspective on digital technology: Understanding the interplay between technology, mind and action. *The Sage handbook of digital technology research*, pages 203–213. Sage, 2013.

Kapur, Manu, and Charles K. Kinzer. Examining the effect of problem type in a synchronous computer-supported collaborative learning (cscl) environment. *Educational Technology Research and Development*, 55(5):439–459, 2007.

Karam, Maria, and M. C. Schraefel. A taxonomy of gestures in human computer interactions. 2005. Retrieved online at: https://eprints.soton.ac.uk/261149/.

Kaye, James A., Maria del Mar Melero-Montes, and Hershel Jick. Mumps, measles, and rubella vaccine and the incidence of autism recorded by general practitioners: A time trend analysis. *BMJ Publishing Group Ltd*, 322(7284):460–463, 2001. DOI: 10.1136/bmj.322.7284.460.

Keifert, Danielle, Christine Lee, Noel Enyedy, Maggie Dahn, Lindsay Lindberg, and Joshua Danish. Tracing bodies through liminal blends in a mixed reality learning environment. *International Journal of Science Education*, 42(18):3093–3115, 2020. DOI: 10.1080/09500693.2020.1851423.

Kendon, Adam. How gestures can become like words. In *Cross-cultural perspectives in nonverbal communication*, pages 131–141. Hogrefe & Huber Publishers, 1988.

Kendon, Adam. Some recent work from italy on "quotable gestures (emblems)." *Journal of Linguistic Anthropology*, 2(1):92–108, 1992.

Kendon, Adam. *Gesture: Visible action as utterance*. Cambridge University Press, 2004.

Kennedy, Jessie B., Kenneth J. Mitchell, and Peter J. Barclay. A framework for information visualisation. *SIGMOD Record*, 25(4):30–34, December 1996. ISSN 0163-5808. DOI: 10.1145/245882.245895.

Kisiel, James, Shawn Rowe, Melanie Ani Vartabedian, and Charles Kopczak. Evidence for family engagement in scientific reasoning at interactive animal exhibits. *Science Education*, 96(6): 1047–1070, 2012.

Klopfer, Eric, Judy Perry, Kurt Squire, Ming-Fong Jan, and Constance Steinkuehler. Mystery at the museum: A collaborative game for museum education. In *Computer Supported Collaborative Learning 2005: The Next 10 Years!*, volume 5, pages 316–320, Routledge, 2005. ISBN 9781351226905

Kong, Ha-Kyung, Zhicheng Liu, and Karrie Karahalios. Frames and slants in titles of visualizations on controversial topics. In *Proceedings of the 2018 CHI Conference on Human Factors in Computing Systems (CHI'18)*, paper 438, pages 1–12, New York, NY, USA, 2018. Association for Computing Machinery. DOI: 10.1145/3173574.3174012.

Köpsel, Anne, and Nikola Bubalo. Benefiting from legacy bias. *Interactions*, 22(5):44–47, 2015.

Krygier, John, and Denis Wood. *Making maps: A visual guide to map design for GIS*. Guilford Publications, 2016.

Lakoff, George. Mapping the brain's metaphor circuitry: Metaphorical thought in everyday reason. *Frontiers in Human Neuroscience*, 8:958, 2014. ISSN 1662-5161. DOI: 10.3389/fnhum.2014.00958.

Lakoff, George. The neural theory of metaphor. *Available at SSRN 1437794*, 2009.

Lakoff, George, Mark Johnson. *Philosophy in the flesh: The embodied mind and its challenge to western thought*, volume 640. Basic Books, 1999.

Lakoff, George, and Mark Johnson. *Metaphors we live by*. University of Chicago Press, 2008.

Lakoff, George, and Rafael Núñez. *Where mathematics comes from*, volume 6. Basic Books, 2000.

Lee, Charlotte P. Boundary negotiating artifacts: Unbinding the routine of boundary objects and embracing chaos in collaborative work. *Computer Supported Cooperative Work (CSCW)*, 16(3): 307–339, 2007.

Lee, Victor R., and Joel Drake. Quantified recess: Design of an activity for elementary students involving analyses of their own movement data. In *Proceedings of the 12th International Conference on Interaction Design and Children (IDC'13)*, pages 273–276, New York, NY, USA, 2013. Association for Computing Machinery. DOI: 10.1145/2485760.2485822.

Lee Ju, Seok, Jongil Lim, Girma Tewolde, and Jaerock Kwon. Autonomous tour guide robot by using ultrasonic range sensors and qr code recognition in indoor environment. In *IEEE International Conference on Electro/Information Technology*, pages 410–415. 2014.

Lee, Sukwon, Sung-Hee Kim, and Bum Chul Kwon. Vlat: Development of a visualization literacy assessment test. *IEEE Transactions on Visualization and Computer Graphics*, 23(1):551–560, 2016.

Lee-Cultura, Serena, Kshitij Sharma, Sofia Papavlasopoulou, Symeon Retalis, and Michail Giannakos. Using sensing technologies to explain children's self-representation in motion-based educational games. In *Proceedings of the Interaction Design and Children Conference (IDC'20)*, pages 541–555, New York, NY, USA, 2020. Association for Computing Machinery. ISBN 9781450379816. DOI: 10.1145/3392063.3394419.

Leinhardt, Gaea, and Kevin Crowley. Conversational elaboration as a process and an outcome of museum learning. *Museum Learning Collaborative Technical Report (MLC-01)*. Learning Research and Development Center, University of Pittsburgh, 1998.

Leinhardt, Gaea, Kevin Crowley, and Karen Knutson. *Learning conversations in museums*. Taylor & Francis, 2003.

Libarkin, Julie C., and Christine Brick. Research methodologies in science education: Visualization and the geosciences. *Journal of Geoscience Education*, 50(4):449–455, 2002.

Lindgren, Robb. *Perspective-based learning in virtual environments*. Stanford University, 2009.

Lindgren, Robb. Generating a learning stance through perspective-taking in a virtual environment. *Computers in Human Behavior*, 28(4):1130–1139, 2012.

Lindgren, Robb, and Mina Johnson-Glenberg. Emboldened by embodiment: Six precepts for research on embodied learning and mixed reality. *Educational Researcher*, 42(8):445–452, 2013. DOI: 10.31020013189X13511661.

Lindgren, Robb, and J. Michael Moshell. Supporting children's learning with body-based metaphors in a mixed reality environment. In *Proceedings of the 10th International Conference on Interaction Design and Children (IDC'11)*, pages 177–180, New York, NY, USA, 2011. Association for Computing Machinery. ISBN 9781450307512. DOI: 10.1145/1999030.1999055.

Liu, Can, Olivier Chapuis, Michel Beaudouin-Lafon, and Eric Lecolinet. Shared interaction on a wall-sized display in a data manipulation task. In *Proceedings of the 2016 CHI Conference on Human Factors in Computing Systems (CHI'16)*, pages 2075–2086, New York, NY, USA, 2016. Association for Computing Machinery. DOI: 10.1145/2858036.2858039.

Lloyd, William J. Integrating gis into the undergraduate learning environment. *Journal of Geography*, 100(5):158–163, 2001.

Locoro, Angela. A fil di dato: valore e comunicazione dell'informazione al tempo dello human-data interaction design. *Mondo Digitale*, volume 17, pages 1–10. Associazione italiana per l'informatica e il calcolo automatico, 2018.

Lozano, Sandra C., Bridgette Martin Hard, and Barbara Tversky. Perspective taking promotes action understanding and learning. *Journal of Experimental Psychology: Human Perception and Performance*, 32(6):1405, 2006. DOI: 10.1037/0096-1523.32.6.1405.

Lyons, Leilah. Designing opportunistic user interfaces to support a collaborative museum exhibit. In *Proceedings of the 9th International Conference on Computer Supported Collaborative Learning—Volume 1*, pages 375–384. International Society of the Learning Sciences, 2009.

Lyons, Leilah, David Becker, and Jessica Roberts. Analyzing the affordances of mobile technologies for informal science learning. *Museums & Social Issues*, 5(1):87–102, 2010.

Lyons, Leilah, Michael Tissenbaum, Matthew Berland, Rebecca Eydt, Lauren Wielgus, and Adam Mechtley. Designing visible engineering: Supporting tinkering performances in museums. In *Proceedings of the 14th International Conference on Interaction Design and Children (IDC'15)*, pages 49–58, New York, NY, USA 2015. Association of Computing Machinery. DOI: 10.1145/2771839.2771845.

Ma, J., I. Liao, K. Ma, and J. Frazier. Living liquid: Design and evaluation of an exploratory visualization tool for museum visitors. *IEEE Transactions on Visualization and Computer Graphics*, 18(12):2799–2808, 2012. DOI: 10.1109/TVCG.2012.244.

Ma, J., K. Ma, and J. Frazier. Decoding a complex visualization in a science museum—An empirical study. *IEEE Transactions on Visualization and Computer Graphics*, 26(1):472–481, 2020.

MacEachren, Alan M. Moving geovisualization toward support for group work. In *Exploring geovisualization*, pages 445–461. Elsevier, 2005.

MacEachren, Alan M., and Isaac Brewer. Developing a conceptual framework for visually-enabled geocollaboration. *International Journal of Geographical Information Science*, 18(1):1–34, 2004.

MacKenzie, I. Scott, and Shawn X. Zhang. The immediate usability of graffiti. In *Proceedings of Graphics Interface '97*, pages 129–137. Canadian Information Processing Society, 1997.

Maher, Mary Lou, and Lina Lee. Designing for gesture and tangible interaction. *Synthesis Lectures on Human-Centered Interaction*, 10(2):i–111, 2017.

Mahon, Bradford Z., and Alfonso Caramazza. A critical look at the embodied cognition hypothesis and a new proposal for grounding conceptual content. *Journal of Physiology-Paris*, 102(1–3): 59–70, 2008.

Malinverni, Laura, and Narcís Parés Burguès. The medium matters: The impact of full-body inter-action on the socio-affective aspects of collaboration. In *Proceedings of the 14th International Conference on Interaction Design and Children (IDC'15)*, pages 89–98, New York, NY, USA, 2015. Association for Computing Machinery. DOI: 10.1145/2771839.2771849.

Marsh, Meredith, Reginald Golledge, and Sarah E. Battersby. Geospatial concept understanding and recognition in g6–college students: A preliminary argument for minimal gis. *Annals of the Association of American Geographers*, 97(4):696–712, 2007.

Marshall, Paul, Richard Morris, Yvonne Rogers, Stefan Kreitmayer, and Matt Davies. Rethink-ing "multi-user" an in-the-wild study of how groups approach a walk-up-and-use tabletop interface. In *Proceedings of the SIGCHI Conference on Human Factors in Computing Systems (CHI'11)*, pages 3033–3042, New York, NY, USA, 2011. Association for Computing Machinery.

Mashhadi, Afra, Fahim Kawsar, and Utku Günay Acer. Human data interaction in iot: The own-ership aspect. In *2014 IEEE World Forum on Internet of Things (WF-IoT)*, pages 159–162. 2014.

McCabe, Craig Andrew. Effects of data complexity and map abstraction on the perception of patterns in infectious disease animations. Masters' thesis. 2009. Retrieved online at: https://etda.libraries.psu.edu/catalog/9949.

McManus, P. M. Families in museums. In *Toward the museum of the future*, 1994.

McNeill, David. *Hand and mind: What gestures reveal about thought*. University of Chicago Press, 1992.

McNeill, David. Gesture: A psycholinguistic approach. *The encyclopedia of language and linguistics*, pages 58–66. Elsevier, 2006.

McNeill, David. *The conceptual basis of language (RLE linguistics A: General linguistics)*. Routledge, 2014.

Merleau-Ponty, Maurice. *Phenomenology of perception*. Motilal Banarsidass Publisher, 1996.

Milgram, Paul, and Fumio Kishino. A taxonomy of mixed reality visual displays. *IEICE TRANS-ACTIONS on Information and Systems*, 77(12):1321–1329, 1994.

Mishra, Swati, and Francesco Cafaro. Full body interaction beyond fun: Engaging museum visitors in human-data interaction. In *Proceedings of the Twelfth International Conference on Tangible, Embedded, and Embodied Interaction (TEI'18)*, pages 313–319, New York, NY, USA, 2018. Association for Computing Machinery. DOI: 10.1145/3173225.3173291.

Mitchell, Kenneth J., Jessie B. Kennedy, and Peter J. Barclay. Using active constructs in user-interfaces to object-oriented databases. In *Proceedings of the 1997 International Conference on International Database Engineering and Applications Symposium (IDEAS'97)*, pages 3–12. IEEE Computer Society, 1997.

Monmonier, Mark. *How to lie with maps*. University of Chicago Press, 2018.

Moore, Leslie M. The basic practice of statistics, *Technometrics*, 38(4):404–405. Taylor and Francis, 1996. DOI: 10.1080/00401706.1996.10484558.

Moran, Dermot. *Introduction to phenomenology*. Routledge, 2002.

Mori, Masahiro, Karl F. MacDorman, and Norri Kageki. The uncanny valley [from the field]. *IEEE Robotics & Automation Magazine*, 19(2):98–100, 2012.

Morris, Meredith Ringel. Web on the wall: Insights from a multimodal interaction elicitation study. In *Proceedings of the 2012 ACM International Conference on Interactive Tabletops and Surfaces (ITS'12)*, pages 95–104, New York, NY, USA, 2012. Association for Computing Machinery. DOI: 10.1145/2396636.2396651.

Morris, Meredith Ringel, Andreea Danielescu, Steven Drucker, Danyel Fisher, Bongshin Lee, M. C. Schraefel, and Jacob O Wobbrock. Reducing legacy bias in gesture elicitation studies. *Interactions*, 21(3):40–45, 2014.

Mortier, Richard, Hamed Haddadi, Tristan Henderson, Derek McAuley, and Jon Crowcroft. *Challenges and opportunities in human-data interaction*. University of Cambridge, Computer Laboratory, 2013.

Mortier, Richard, Hamed Haddadi, Tristan Henderson, Derek McAuley, and Jon Crowcroft. Human-data interaction: The human face of the data-driven society. *Available at SSRN 2508051*, 2014.

Müller, Jörg, Robert Walter, Gilles Bailly, Michael Nischt, and Florian Alt. Looking glass: A field study on noticing interactivity of a shop window. In *Proceedings of the SIGCHI Conference on Human Factors in Computing Systems (CHI'12)*, pages 297–306, New York, NY, USA, 2012. Association for Computing Machinery.

Murray, Scott T., Irwin S. Kirsch, and Lynn Jenkins. *Adult Literacy in OECD Countries: Technical Report on the First International Adult Literacy Survey*. U.S. Department of Education, Office of Educational Research and Improvement, 1998.

Nacenta, Miguel A., Yemliha Kamber, Yizhou Qiang, and Per Ola Kristensson. Memorability of pre-designed and user-defined gesture sets. In *Proceedings of the SIGCHI Conference on Human Factors in Computing Systems (CHI'13)*, pages 1099–1108, New York, NY, USA, 2013. Association for Computing Machinery.

Narechania, Arpit, Arjun Srinivasan, and John Stasko. Nl4dv: A toolkit for generating analytic specifications for data visualization from natural language queries. *IEEE Transactions on Visualization and Computer Graphics*, 27(2):369–379, 2021. DOI: 10.1109/TVCG.2020.3030378.

National Research Council. *Learning science in informal environments: People, places, and pursuits*. National Academies Press, 2009.

Neininger, Bettina, and Friedemann Pulvermüller. Word-category specific deficits after lesions in the right hemisphere. *Neuropsychologia*, 41(1):53–70, 2003.

Norman, Donald A. Natural user interfaces are not natural. *Interactions*, 17(3):6–10, 2010.

Norman, Donald A., and Jakob Nielsen. Gestural interfaces: A step backward in usability. *Interactions*, 17(5):46–49, 2010.

Novella, S. Evidence in medicine: Correlation and causation. *Science-Based Medicine*. 2009. Retrieved online at: https://sciencebasedmedicine.org/evidence-in-medicine-correlation-and-causation/.

Ochs, Elinor, Patrick Gonzales, and Sally Jacoby. "When i come down i'm in the domain state": Grammar and graphic representation in the interpretive activity of physicists. *Studies in Interactional Sociolinguistics*, 13:328–369, 1996.

Ojala, T., V. Kostakos, H. Kukka, T. Heikkinen, T. Linden, M. Jurmu, S. Hosio, F. Kruger, and D. Zanni. Multipurpose interactive public displays in the wild: Three years later. *Computer*, 45(5):42–49, May 2012. ISSN 00189162. DOI: 10.1109/MC.2012.115.

Packer, Jan, and Roy Ballantyne. Solitary vs. shared: Exploring the social dimension of museum learning. *Curator: The Museum Journal*, 48(2):177–192, 2005.

Papert, Seymour. *Mindstorms: Children, computers, and powerful ideas*. Basic Books, 2020.

Papert, Seymour, and Idit Harel. Situating constructionism. *Constructionism*, 36(2):1–11, 1991.

Paris, Scott G. Situated motivation and informal learning. *Journal of Museum Education*, 22(2–3): 22–27, 1997.

Pearson, K. *The Grammar of Science*. White, 1911.

Peck, Evan M., Sofia E. Ayuso, and Omar El-Etr. Data is personal: Attitudes and perceptions of data visualization in rural pennsylvania. In *Proceedings of the 2019 CHI Conference on Human Factors in Computing Systems*, paper 244, pages 1–12, New York, NY, USA, 2019. Association for Computing Machinery. DOI: 10.1145/3290605.3300474.

Perer, Adam, and Ben Shneiderman. Systematic yet flexible discovery: Guiding domain experts through exploratory data analysis. In *Proceedings of the 13th International Conference on Intelligent User Interfaces (IUI'08)*, pages 109–118, New York, NY, USA, 2008. Association for Computing Machinery. ISBN 9781595939876. DOI: 10.1145/1378773.1378788.

Pérez-Sanagustín, Mar, Denis Parra, Renato Verdugo, Gonzalo García-Galleguillos, and Miguel Nussbaum. Using qr codes to increase user engagement in museum-like spaces. *Computers in Human Behavior*, 60:73–85, 2016.

Perry, Deborah L. Beyond cognition and affect: The anatomy of a museum visit. In *Visitor studies: Theory, research and practice: Collected papers from the 1993 Visitor Studies Conference*, volume 6, pages 43–47. Taylor and Francis, 1993.

Perry, Deborah L. *What makes learning fun? Principles for the design of intrinsically motivating museum exhibits*. Rowman Altamira, 2012.

Petrelli, Daniela, and Sinead O'Brien. Phone vs. tangible in museums: A comparative study. In *Proceedings of the 2018 CHI Conference on Human Factors in Computing Systems (CHI'18)*, New

York, NY, USA, 2018. Association for Computing Machinery. ISBN 9781450356206. DOI: 10.1145/3173574.3173686.

Pillat, Remo, Arjun Nagendran, and Robb Lindgren. A mixed reality system for teaching stem content using embodied learning and whole-body metaphors. In *Proceedings of the 11th ACM SIGGRAPH International Conference on Virtual-Reality Continuum and Its Applications in Industry (VRCAI'12)*, pages 295–302, New York, NY, USA, 2012. Association for Computing Machinery. ISBN 9781450318259. DOI: 10.1145/2407516.2407584.

Piumsomboon, Thammathip, Adrian Clark, Mark Billinghurst, and Andy Cockburn. User-defined gestures for augmented reality. In *IFIP Conference on Human-Computer Interaction*, pages 282–299. Springer, 2013.

Plass, Jan L., Bruce D. Homer, and Elizabeth O. Hayward. Design factors for educationally effective animations and simulations. *Journal of Computing in Higher Education*, 21(1):31–61, 2009.

Pousman, Zachary, John Stasko, and Michael Mateas. Casual information visualization: Depictions of data in everyday life. *IEEE Transactions on Visualization and Computer Graphics*, 13(6): 1145–1152, 2007. DOI: 10.1109/TVCG.2007.70541.

Povis, Kaleen Tison, and Kevin Crowley. Family learning in object-based museums: The role of joint attention. *Visitor Studies*, 18(2):168–182, 2015.

Prante, Thorsten, Carsten Röcker, Norbert Streitz, Richard Stenzel, Carsten Magerkurth, Daniel Van Alphen, and Daniela Plewe. Hello. Wall—Beyond ambient displays. In *Adjunct Proceedings of the Fifth International Conference on Ubiquitous Computing (UBICOMP'03)*, pages 277–278, Seattle, WA, USA, 2003.

Price, Sara, Mona Sakr, and Carey Jewitt. Exploring whole-body interaction and design for museums. *Interacting with Computers*, 28(5):569–583, 2016.

Prouzeau, Arnaud, Anastasia Bezerianos, and Olivier Chapuis. Evaluating multi-user selection for exploring graph topology on wall-displays. *IEEE Transactions on Visualization and Computer Graphics*, 23(8):1936–1951, 2016. DOI: 10.1109/TVCG.2016.2592906.

Radinsky, J., J. Melendez, and J. Roberts. Do the data strike back? Students' presentations of historical narratives about latino communities using gis. In Josh Radinsky (Chair), *Tools for Constructing Historical Narratives: Teaching African American and Latino Histories With GIS Census Maps*. Symposium conducted at the meeting of the American Educational Research Association, Vancouver, BC, Canada, 2012.

Ress, S., Francesco Cafaro, D. Bora, D. Prasad, and D. Soundarajan. Mapping history: Orienting museum visitors across time and space. *Journal on Computing and Cultural Heritage (JOCCH)*, 11(3):1–25, 2018.

Richards, Ted. Using kinesthetic activities to teach ptolemaic and copernican retrograde motion. *Science & Education*, 21(6):899–910, 2012.

Rittschof, Kent A., and Raymond W Kulhavy. Learning and remembering from thematic maps of familiar regions. *Educational Technology Research and Development*, 46(1):19–38, 1998.

Roberts, Jessica, L. Lyons, and J. Radinsky. Become one with the data: Technological support of shared exploration of data in informal settings. In Anne Knowles (Chair), *From Visualizing to Understanding Historical Change: Using GIS Tools on the Web, in Class, and in Museums*. Paper session conducted at the meeting of the Social Science History Association, Chicago, IL, 2013a.

Roberts, Jessica, Amartya Banerjee, Annette Hong, Steven McGee, Michael Horn, and Matt Matcuk. Digital exhibit labels in museums: Promoting visitor engagement with cultural artifacts. In *Proceedings of the 2018 CHI Conference on Human Factors in Computing Systems (CHI'18)*, pages 1–12, New York, NY, USA, 2018. Association for Computing Machinery.

Roberts, Jessica, Francesco Cafaro, Raymond Kang, Kristen Vogt, Leilah Lyons, and Josh Radinsky. That's me and that's you: Museum visitors' perspective-taking around an embodied interaction data map display. In To *see the world and a grain of sand: Learning across levels of space, time, and scale*. CSCL 2013 Conference Proceedings, 2:343–344. Madison, WI: International Society of the Learning Sciences, 2013b.

Roberts, Jessica, L. Lyons, F. Cafaro, and R. Eydt. Harnessing motion-sensing technologies to engage visitors with digital data. In *Proceedings of Museums and the Web*, 2015. Retrieved online at: https://mw2015.museumsandtheweb.com/paper/harnessing -sensing-technologies-to-engage-visitors-with-digital-data/.

Roberts, Jessica, and Leilah Lyons. Examining spontaneous perspective taking and fluid self-to-data relationships in informal open-ended data exploration. *Journal of the Learning Sciences*, 29(1):32–56, 2020.

Roberts, Jessica, and Leilah Lyons. Scoring qualitative informal learning dialogue: The squild method for measuring museum learn phia, PA: International Society of the Learning Sciences., 2017a.

Roberts, Jessica, and Leilah Lyons. The value of learning talk: Applying a novel dialogue scoring method to inform interaction design in an open-ended, embodied museum exhibit. *International Journal of Computer-Supported Collaborative Learning*, 12(4):343–376, 2017b.

Roberts, Jessica, Leilah Lyons, Francesco Cafaro, and Rebecca Eydt. Interpreting data from within: Supporting humandata interaction in museum exhibits through perspective taking. In *Proceedings of the 2014 conference on Interaction design and children (IDC'14)*, pages 7–16, New York, NY, USA, 2014. Association for Computing Machinery. DOI: 10.1145/2593968.2593974.

Roberts, Jessica, Josh Radinsky, Leilah Lyons, and Francesco Cafaro. Co-census: Designing an interactive museum space to prompt negotiated narratives of ethnicity, community, and identity. Paper presented at the 2012 meeting of the *American Educational Reserach Association*, Vancouver, BC, Canada, 2012. Retrieved online at: https://cocensus.uic.edu/wp-content/uploads/2014/10/AERA_2012.pdf.

Roberts, L. *From knowledge to narrative*. Smithsonian Institute, 1997.

Robinson, Edward Stevens, Irene Case Sierman, and Lois E. Curry. *The behavior of the museum visitor*. 1928. Retrieved online at: https://files.eric.ed.gov/fulltext/ED044919.pdf.

Rogers, Yvonne. Hci theory: Classical, modern, and contemporary. *Synthesis Lectures on Human-Centered Informatics*, 5(2):1–129, 2012.

Rogers, Yvonne, Helen Sharp, and Jenny Preece. *Interaction design: Beyond human-computer interaction*. John Wiley & Sons, 2011.

Rounds, Jay. Doing identity work in museums. *Curator: The Museum Journal*, 49(2):133–150, 2006.

Rowe, Shawn M., James V. Wertsch, and Tatyana Y. Kosyaeva. Linking little narratives to big ones: Narrative and public memory in history museums. *Culture & Psychology*, 8(1):96–112, 2002.

Ruiz, Jaime, Yang Li, and Edward Lank. User-defined motion gestures for mobile interaction. In *Proceedings of the SIGCHI Conference on Human Factors in Computing Systems (CHI'11)*, pages 197–206, New York, NY, USA, 2011. Association for Computing Machinery. ISBN 9781450302289. DOI: 10.1145/1978942.1978971.

Russell, Bertrand. On the notion of cause. In *Proceedings of the Aristotelian Society*, volume 13, pages 1–26. Oxford University Press, 1912.

Sailaja, Neelima, Rhianne Jones, and Derek McAuley. Designing for human data interaction in data-driven media experiences. In *Extended Abstracts of the 2021 CHI Conference on Human Factors in Computing Systems*, CHI EA '21, New York, NY, USA, 2021. Association for Computing Machinery. ISBN 9781450380959. DOI: 10.1145/3411763.3451808.

Saldaña, Johnny. *The coding manual for qualitative researchers*. Sage, 2015.

Sandifer, Cody. Time-based behaviors at an interactive science museum: Exploring the differences between weekday/weekend and family/nonfamily visitors. *Science Education*, 81(6):689–701, 1997.

Sawicki, David S., and William J. Craig. The democratization of data: Bridging the gap for community groups. *Journal of the American Planning Association*, 62(4):512–523, 1996.

Schauble, Leona, Mary Gleason, Rich Lehrer, Karol Bartlett, Anthony Petrosino, Annie Allen, and J. Street. Supporting science learning in museums. In *Learning conversations in museums*, 425–452. Lawrence Erlbaum Associates Publishers, 2002.

Schiano, Diane J., and Barbara Tversky. Structure and strategy in encoding simplified graphs. *Memory & Cognition*, 20(1):12–20, 1992. DOI: 10.3758/bf03208249.

Schönauer, Christian, Thomas Pintaric, and Hannes Kaufmann. Full body interaction for serious games in motor rehabilitation. In *Proceedings of the 2nd Augmented Human International Conference (AH'11)*, New York, NY, USA, 2011. Association for Computing Machinery. ISBN 9781450304269. DOI: 10.1145/1959826.1959830.

Schultz, Michelle Kelly. A case study on the appropriateness of using quick response (qr) codes in libraries and museums. *Library & Information Science Research*, 35(3):207–215, 2013.

Segal, Ayelet. *Do gestural interfaces promote thinking? Embodied interaction: Congruent gestures and direct-touch promote performance in math*. Ph.D. thesis, Columbia University, 2011.

Shadish, William R., Thomas D. Cook, Donald Thomas Campbell. *Experimental and quasi-experimental designs for generalized causal inference*. Houghton Mifflin, 2002.

Shah, Priti, and James Hoeffner. Review of graph comprehension research: Implications for instruction. *Educational Psychology Review*, 14(1):47–69, 2002.

Shapiro, Ben Rydal, Rogers P. Hall, and David A. Owens. Developing and using interaction geography in a museum. *International Journal of Computer-Supported Collaborative Learning*, 12 (4):377–399, 2017.

Shapiro, Ben Rydal, and Rogers Hall. Personal curation in a museum. *Proceedings of the ACM on Human-Computer Interaction*, 2(CSCW):1–22, 2018.

Shapiro, Lawrence. *Embodied cognition*. Routledge, 2010. ISBN 1136963936.

Singer, Melissa, Joshua Radinsky, and Susan R. Goldman. The role of gesture in meaning construction. *Discourse Processes*, 45(4–5):365–386, 2008.

Snibbe, Scott S., and Hayes S. Raffle. Social immersive media: Pursuing best practices for multi-user interactive camera/projector exhibits. In *Proceedings of the SIGCHI Conference on Human Factors in Computing Systems (CHI'09)*, pages 1447–1456, New York, NY, USA, 2009. Association for Computing Machinery. DOI: 10.1145/1518701.1518920.

Soni, Nikita, Alice Darrow, Annie Luc, Schuyler Gleaves, Carrie Schuman, Hannah Neff, Peter Chang, Brittani Kirkland, Jeremy Alexandre, Amanda Morales, Kathryn A. Stofer, and Lisa Anthony. Affording embodied cognition through touchscreen and above-the-surface gestures during collaborative tabletop science learning. *International Journal of Computer-Supported Collaborative Learning*, 16:1–40, 2021.

Sprague, David, and Melanie Tory. Exploring how and why people use visualizations in casual contexts: Modeling user goals and regulated motivations. *Information Visualization*, 11(2): 106–123, 2012.

Stahl, Gerry. A model of collaborative knowledge-building. In *Fourth international conference of the learning sciences*, volume 10, pages 70–77. Lawrence Erlbaum Associates, 2000.

Star, Susan Leigh, and James R. Griesemer. Institutional ecology, "Translations" and boundary objects: Amateurs and professionals in berkeley's museum of vertebrate zoology, 1907-39. *Social Studies of Science*, 19(3):387–420, 1989.

Strauss, Anselm, and Juliet M. Corbin. *Grounded theory in practice*. Sage, 1997.

Suppes, Patrick. A probabilistic theory of causality. *Philosophy of Science* (The University of Chicago Press), 1972.

Szafir, Danielle Albers. Modeling color difference for visualization design. *IEEE Transactions on Visualization and Computer Graphics*, 24(1):392–401, 2017. DOI: 10.1109/TVCG.2017 .2744359.

Tashakkori, Abbas, and Charles Teddlie. *Sage handbook of mixed methods in social & behavioral research*. Sage, 2010.

Thian, Cherry. Augmented reality—What reality can we learn from it. In *Museums and the Web*, 2012. Retrieved online at: https://www.museumsandtheweb.com/mw2012/papers/augmented _reality_what_reality_can_we_learn_fr.

Thomas, J. J., and K. A. Cook. A visual analytics agenda. *IEEE Computer Graphics and Applications*, 26(1):10–13, 2006. DOI: 10.1109/MCG.2006.5.

Trajkova, Milka, A'aeshah Alhakamy, Francesco Cafaro, Rashmi Mallappa, and Sreekanth R. Kankara. Move your body: Engaging museum visitors with human-data interaction. In *Proceedings of the 2020 CHI Conference on Human Factors in Computing Systems (CHI'20)*, pages 1–13, New York, NY, USA, 2020a. Association for Computing Machinery. ISBN 9781450367080. DOI: 10.1145/3313831.3376186.

Trajkova, Milka, Francesco Cafaro, Sanika Vedak, Rashmi Mallappa, Sreekanth R. Kankara. Exploring casual covid-19 data visualizations on twitter: Topics and challenges. In *Informatics*, volume 7, page 35. Multidisciplinary Digital Publishing Institute, 2020b.

Tukey, John W. (John Wilder). *Exploratory data analysis*. Addison-Wesley series in behavioral science. Addison-Wesley Pub. Co., 1977. ISBN 0201076160.

Uttal, David H. Seeing the big picture: Map use and the development of spatial cognition. *Developmental Science*, 3(3):247–264, 2000.

Van Dam, Andries. Post-wimp user interfaces. *Communications of the ACM*, 40(2):63–67, 1997.

Van Dam, Laura. A picture is worth 1,000 numbers. *Technology Review*, 95(4):34–34, 1992.

Vatavu, Radu-Daniel, Gabriel Cramariuc, and Doina Maria Schipor. Touch interaction for children aged 3 to 6 years: Experimental findings and relationship to motor skills. *International Journal of Human-Computer Studies*, 74:54–76, 2015.

Vatavu, Radu-Daniel, and Jacob O. Wobbrock. Formalizing agreement analysis for elicitation studies: New measures, significance test, and toolkit. In *Proceedings of the 33rd Annual ACM Conference on Human Factors in Computing Systems (CHI'15)*, pages 1325–1334, New York,

NY, USA, 2015. Association for Computing Machinery. ISBN 9781450331456. DOI: 10.1145/2702123.2702223.

Vatavu, Radu-Daniel, and Ionut-Alexandru Zaiti. Leap gestures for tv: Insights from an elicitation study. In *Proceedings of the ACM International Conference on Interactive Experiences for TV and Online Video (TVX'14)*, pages 131–138, 2014. Association for Computing Machinery. DOI: 10.1145/2602299.2602316.

Véron, Eliséo, and Martine Levasseur. *Ethnographie de l'exposition: l'espace, le corps et le sens*. Centre Georges Pompidou, Bibliothèque publique d'information, 1989.

Victorelli, Eliane Zambon, Julio Cesar Dos Reis, Heiko Hornung, and Alysson Bolognesi Prado. Understanding human-data interaction: Literature review and recommendations for design. *International Journal of Human-Computer Studies*, 134:13–32, 2020. ISSN 10715819. Retrieved online at: https://doi.org/10.1016/j.ijhcs.2019.09.004.

Villarreal-Narvaez, Santiago, Jean Vanderdonckt, Radu-Daniel Vatavu, and Jacob O. Wobbrock. A systematic review of gesture elicitation studies: What can we learn from 216 studies? In *Proceedings of the 2020 ACM Designing Interactive Systems Conference (DIS'20)*, pages 855–872, New York, NY, USA, 2020. Association for Computing Machinery. ISBN 9781450369749. DOI: 10.1145/3357236.3395511.

Vom Lehn, Dirk, Christian Heath, and Jon Hindmarsh. Exhibiting interaction: Conduct and collaboration in museums and galleries. *Symbolic Interaction*, 24(2):189–216, 2001.

Vygotsky, Lev S. Mind in society: The development of higher psychological processes. Harvard University Press, 1980.

Wachs, Juan Pablo, Mathias Kölsch, Helman Stern, and Yael Edan. Vision-based hand-gesture applications. *Communications of the ACM*, 54(2):60–71, 2011.

Wall, Emily, John Stasko, and Alex Endert. Toward a design space for mitigating cognitive bias in vis. In *2019 IEEE Visualization Conference (VIS)*, pages 111–115. 2019.

Wang, Miaosen, Sebastian Boring, and Saul Greenberg. Proxemic peddler: A public advertising display that captures and preserves the attention of a passerby. In *Proceedings of the 2012 International Symposium on Pervasive Displays (PerDis'12)*, pages 3:1–3:6, New York, NY, USA, 2012. Association for Computing Machinery. ISBN 9781450314145. DOI: 10.1145/2307798.2307801.

Wang, Q., G. Kurillo, F. Ofli, and R. Bajcsy. Evaluation of pose tracking accuracy in the first and second generations of microsoft kinect. In *2015 International Conference on Healthcare Informatics*, pages 380–389, 2015. DOI: 10.1109/ICHI.2015.54.

Watson, John B. *Behaviorism*. Routledge, 2017.

Weiser, Mark. The computer for the 21st century. *Scientific American*, 265(3):94–105, 1991.

Wells, Gordon. Dialogic inquiry in education. *Vygotskian perspectives on literacy research*, pages 51–85. Cambridge University Press, 2000.

Wertsch, James V. *Mind as action*. Oxford University Press, 1998.

Williams, Adam S., Jason Garcia, Fernando De Zayas, Fidel Hernandez, Julia Sharp, and Francisco R Ortega. The cost of production in elicitation studies and the legacy bias-consensus trade off. *Multimodal Technologies and Interaction*, 4(4):88, 2020.

Williams, Amanda, Eric Kabisch, and Paul Dourish. From interaction to participation: Configuring space through embodied interaction. In *International Conference on Ubiquitous Computing*, pages 287–304. Springer, 2005.

Wilson, Robert A., and Lucia Foglia. *Embodied cognition*. Springer, 2011.

Wilson, Margaret. Six views of embodied cognition. *Psychonomic Bulletin & Review*, 9(4):625–636, 2002. ISSN 1069-9384. DOI: 10.3758/BF03196322.

Wobbrock, Jacob O. A robust design for accessible text entry. *ACM SIGACCESS Accessibility and Computing*, (84):48–51, 2006.

Wobbrock, Jacob O., Htet Htet Aung, Brandon Rothrock, and Brad A. Myers. Maximizing the guessability of symbolic input. In *CHI '05 Extended Abstracts on Human Factors in Computing Systems (CHI EA'05)*, pages 1869–1872, New York, NY, USA, 2005. Association for Computing Machinery. ISBN 1595930027. DOI: 10.1145/1056808.1057043.

Wobbrock, Jacob O., Meredith Ringel Morris, and Andrew D. Wilson. User-defined gestures for surface computing. In *Proceedings of the SIGCHI Conference on Human Factors in Computing Systems (CHI'09)*, pages 1083–1092, New York, NY, USA, 2009. Association for Computing Machinery. ISBN 9781605582467. DOI: 10.1145/1518701.1518866.

Wolcott, Donna L., and Thomas G. Wolcott. High mortality of piping plovers on beaches with abundant ghost crabs: Correlation, not causation. *The Wilson Bulletin*, 111(3):321–329, 1999.

Wong, Pak Chung, and Jim Thomas. Visual analytics. *IEEE Computer Graphics and Applications*, (5):20–21, 2004.

Xiong, Cindy, Joel Shapiro, Jessica Hullman, and Steven Franconeri. Illusion of causality in visualized data. *IEEE Transactions on Visualization and Computer Graphics*, 26(1):853–862, 2020. DOI: 10.1109/TVCG.2019.2934399.

Yatani, Koji, Mayumi Onuma, Masanori Sugimoto, and Fusako Kusunoki. Musex: A system for supporting children's collaborative learning in a museum with pdas. *Systems and Computers in Japan*, 35(14):54–63, 2004.

Yiannoutsou, Nikoleta, Ioanna Papadimitriou, Vassilis Komis, and Nikolaos Avouris. "Playing with" museum exhibits: Designing educational games mediated by mobile technology. In *Proceedings of the 8th International Conference on Interaction Design and Children*, pages 230–233, 2009.

Zalta, Edward N., Uri Nodelman, Colin Allen, and John Perry. *Stanford encyclopedia of philosophy*, 1995. Retrieved online at: https://plato.stanford.edu/.

Zhu, Yitan, Huai Li, David J Miller, Zuyi Wang, Jianhua Xuan, Robert Clarke, Eric P. Hoffman, and Yue Wang. Cabig visda: Modeling, visualization, and discovery for cluster analysis of genomic data. *BMC Bioinformatics*, 9(1):383, 2008. DOI: 10.1186/1471-2105-9-383.

Zimmerman, Heather Toomey, Suzanne Reeve, and Philip Bell. Distributed expertise in a science center: Social and intellectual role-taking by families. *Journal of Museum Education*, 33(2): 143–152, 2008.

Authors' Biographies

Francesco Cafaro is an assistant professor in the Department of Human-Centered Computing, School of Informatics and Computing at Indiana University-Purdue University Indianapolis (IUPUI). His work is deeply multi-disciplinary and investigates how theories from learning, cognitive, and computer sciences can provide the scaffolding for the design of embodied interaction. He has led the design and implementation of interactive data visualizations that have been tested at the Jane Addams Hull House in Chicago, the New York Hall of Science in Queens, Historic New Harmony in Indiana, and Discovery Place in Charlotte, NC.

Jessica Roberts is an assistant professor in the School of Interactive Computing at Georgia Tech. She holds a Ph.D. in the Learning Sciences from the University of Illinois-Chicago with a concentration in geospatial analysis and visualization and a B.S. from Northwestern University with a concentration in theatre design. Her research focuses on public engagement with science, with an emphasis on how people learn through, with, and about data in out-of-school environments such as museums and citizen science and how interactive technologies mediate social, informal learning experiences. Her work on the design of interactive learning technologies has won paper awards at CSCL and CHI, and her projects have been exhibited at venues including the Field Museum of Natural History in Chicago and the New York Hall of Science.

Printed in the United States
by Baker & Taylor Publisher Services